Holt Science and Technology
Grade 7

Tennessee Comprehensive Assessment Program Test Preparation Workbook

Copyright © Holt McDougal, a division of Houghton Mifflin Harcourt Publishing Company. All rights reserved.

Warning: No part of this publication may be reproduced or transmitted in any form or by any means, electronic or mechanical, including photocopy, and recording, or by any information storage or retrieval system without the prior written permission of Holt McDougal unless such copying is expressly permitted by federal copyright law. Requests for permission to make copies of any part of the work should be mailed to the following address: Permissions Department, Holt McDougal, 10801 N. MoPac Expressway, Building 3, Austin, Texas 78759.

Teachers using HOLT SCIENCE & TECHNOLOGY may photocopy complete pages in sufficient quantities for classroom use only and not for resale.

HOLT MCDOUGAL is a trademark of Houghton Mifflin Harcourt Publishing Company.

Printed in the United States of America

If you have received these materials as examination copies free of charge, Holt, Rinehart and Winston retains title to the materials and they may not be resold. Resale of examination copies is strictly prohibited.

Possession of this publication in print format does not entitle users to convert this publication, or any portion of it, into electronic format.

ISBN 13: 978-0-55-401785-3
ISBN 10: 0-55-401785-7

5 6 7 8 9 10 11 12 13 0868 15 14 13 12 11
4500315752

Contents

Copyright © by Holt McDougal. All rights reserved.

Introduction

This workbook consists of practice activities designed to prepare your students to take the *Tennessee Comprehensive Assessment Program Test*. The questions are correlated to the *Tennessee Science Standards* as well as the appropriate Skills and Processes Standards. This breadth of content coverage provides teachers with an opportunity to assess their students' understanding of the essential science knowledge and skills at the middle school level. These assessments can then help identify topics or concepts in need of re-teaching or additional practice and should be used to inform curricular decisions on the classroom or school levels.

Copyright © by Holt McDougal. All rights reserved.

Grade 7 – Inquiry

GLE 0707.Inq.1 Design and conduct open-ended scientific investigations.

STANDARD REVIEW

Designing a good experiment requires planning. Every factor should be considered. For example, a possible hypothesis about deformities in frogs is that they were exposed to an increased amount of ultraviolet (UV) light as eggs. If an increase in exposure to ultraviolet light is causing the deformities, then some frog eggs exposed to increasing amounts of ultraviolet light in a laboratory will develop into deformed frogs.

An experiment to test this hypothesis is summarized in Table 1. In this case, the variable is the length of time the eggs are exposed to UV light. All other factors, such as the temperature of the water, are the same in the control group and in the experimental groups.

In a well-designed experiment, the differences between control and experimental groups are caused by the variable and not by differences between individuals. The larger the groups are, the smaller the effect of a difference between individual frogs will be. The larger the groups are, the more likely it is that the variable is responsible for any changes and the more accurate the data collected are likely to be. Scientists test a result by repeating the experiment. If an experiment gives the same results each time, scientists are more certain about the variable's effect on the outcome.

Table 1 Experiment to Test Effect of UV Light on Frogs

Group	Control Factors			Variable
	Kind of frog	Number of eggs	Temperature	UV light exposure
1 (Control)	Leopard frog	100	25°C	0 days
2 (Experimental)	Leopard frog	100	25°C	15 days
3 (Experimental)	Leopard frog	100	25°C	24 days

GUIDED PRACTICE

Directions: Using the Standard Review and what you have studied, read each question and circle the letter of the best response.

Which choice below best describes a hypothesis?

A the result of an experiment

B a well-tested explanation for why something happens

C a random guess about what causes something

D a possible explanation of an observation or resul

The correct answer is D. A hypothesis is a possible explanation based on observations or reason, so it is not a random guess (Answer C). It is the basis for an experiment, not a result (Answer A). A well-tested explanation (Answer B) is a theory, not a hypothesis.

Copyright © Holt McDougal. All rights reserved.

STANDARD PRACTICE

1. Lamont conducted an investigation of four ant piles to see whether the amount of sunlight they got affected their activity. One ant pile was located in a sunny area, one ant pile was in a shady area, and the other two ant piles were in partly shaded areas. After five days, he noticed that the ant pile in the shady area had no ants. Which of the following statements best describes what Lamont should do?

A He should stop observing that ant pile because it no longer pertains to his investigation.

B He should revise his hypothesis and begin a new investigation omitting the shaded ant pile.

C He should make a note of the date and time that he noticed no activity in the shaded ant pile.

D He should remove the object that is shading that ant pile so that it will receive more sunlight.

2. What is an observation?

F a test performed on psychological subjects

G any information you gather through your senses

H the set of tools and methods used to collect data in science

J a procedure that makes use of scientific equipment to record information

3. What makes a hypothesis testable?

A if it is based on accurate observations

B if an experiment can be designed to test it

C if it contains an opinion that other scientists share

D if it is based on a theory accepted by many scientists

4. a. Why do hypotheses need to be testable?

b. A scientist who studies mice observes that on the day the mice are fed vitamins with their meals, they perform better in mazes. What hypothesis would you form to explain this phenomenon?

Copyright © Holt McDougal. All rights reserved.

GLE 0707.Inq.2 Use appropriate tools and techniques to gather, organize, analyze, and interpret data.

STANDARD REVIEW

In most experiments, you will measure some quantity, such as distance, mass, temperature, or time. When you work in the lab, you will use scientific apparatus to make these measurements and control experiments. It is important that you follow the correct procedures for using the apparatus. Techniques for using equipment will be provided by your teacher or in instructions in your lab book. It is important to read and follow all procedures exactly. If you use equipment incorrectly, you may create a safety hazard or get incorrect results. It is important to follow instructions in the correct sequence and observe all safety guidelines.

After you finish collecting data, you will use tables, graphs, diagrams, maps, or other visual displays, to analyze and interpret it. These tools point out the relationships in the numbers that record your measurements. Part of conducting a successful experiment is analyzing your data to find any hidden patterns. Two common patterns that you might see on a graph of experimental data are *linear relationships,* in which data tend to form a straight line, and *repeating relationships,* in which a cycle of changing values shows up in the analysis. When you analyze data, you also need to look at methods of measurement and calculation to find the difference between a predicted amount and a calculated amount.

GUIDED PRACTICE

Directions: Using the Standard Review and what you have studied, read each question and circle the letter of the best response.

What should you do if you are not sure how to use a piece of laboratory equipment?

A Stop working on the experiment.

B Ask another student to show you how to use the equipment.

C Check the lab instructions for the correct techniques.

D Make your best guess about how it works.

The correct answer is C. When you run an experiment that uses equipment, you can get information on using it from the lab instructions or from your teacher. You should find out how to use the equipment, and then continue the experiment (Answer A). You should never guess if you don't know how to use the equipment (Answer D) or ask someone other than your teacher (Answer B) because you might introduce an error that could cause a safety hazard or change the results of the experiment.

Copyright © Holt McDougal. All rights reserved.

Grade 7 – Inquiry

STANDARD PRACTICE

1. Why might a scientist use colored dyes when viewing items under a compound light microscope?

A to make them more visible

B to make them more attractive

C to make them appear more realistic

D to make them appear more three-dimensional

2. Where should you take the reading when you are measuring a liquid in a glass graduated cylinder?

F It does not matter as long as you use the same method for every measurement.

G Read the mark closest to the center of the curve in the liquid's surface.

H Read the mark closest to the very edge where the liquid touches the sides of the graduated cylinder.

J Read the mark closest to the point halfway between the sides of the graduated cylinder and the center of the curve in the liquid's surface.

3. Which of the following sources of information would be the least likely place to look for good scientific information about nutrition?

A www.nih.gov

B www.ucla.edu

C www.sciencenews.org

D www.nutritionbarz.com

4. a. Why is it important to question critically scientific information found on an unfamiliar Web site?

b. Give an example of how a life scientist might use computers and technology.

Copyright © Holt McDougal. All rights reserved.

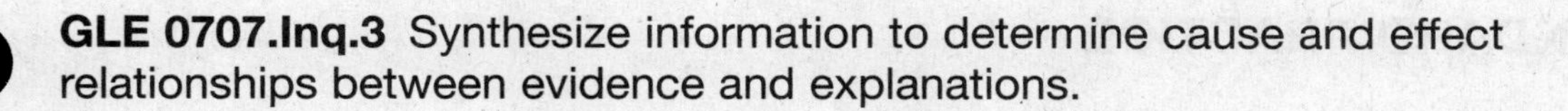

Grade 7 – Inquiry

GLE 0707.Inq.3 Synthesize information to determine cause and effect relationships between evidence and explanations.

STANDARD REVIEW

Once scientists finish their tests, they must analyze the results. Scientists often make graphs and tables to organize and summarize their data. Analysis also includes comparing new data with known information to discover what new information the new data provides. After carefully analyzing the results of their tests, scientists must decide whether their results supported the hypothesis. The conclusion is an interpretation of the results and how they compare to the original ideas of the hypothesis. If a scientist concludes that the results support the original hypothesis, the conclusion may suggest new questions for further study.

One way to test a hypothesis is to do a controlled experiment. A *controlled experiment* compares the results from a control group with the results from one or more experimental groups. The control group and the experimental groups are the same except for one factor. This factor is called a *variable*. The experiment will then show the effect of the variable. If your experiment has more than one variable, determining which variable is responsible for the experiment's results will be difficult or impossible.

Pressure measured in a closed container

Volume	Temperature	Gas in Container	Pressure
100 mL	0°C	nitrogen	100 kPa
100 mL	25°C	nitrogen	109 kPa
100 mL	50°C	nitrogen	118 kPa

GUIDED PRACTICE

Directions: Using the Standard Review and what you have studied, read each question and circle the letter of the best response.

In an experiment, a closed container is heated, and the pressure inside the container is measured. Which of these conclusions can you make based on the data below?

A Increasing the volume causes an increase in pressure.

B Increasing the temperature of nitrogen in a closed container causes the pressure to increase.

C Increasing the temperature of any material causes its pressure to increase.

D Increasing the pressure in a closed container causes the temperature to increase.

The correct answer is B. Answer D is incorrect because the experiment changed the temperature and observed the pressure. Answer A is incorrect because the volume was constant. Because only nitrogen was studied, you cannot make conclusions about other materials (Answer C).

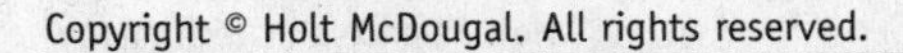

Copyright © Holt McDougal. All rights reserved.

STANDARD PRACTICE

1. A dock was built over a large bed of sea grass in a manatee habitat. The dock shaded the bed of sea grass from the sun. The population of manatees decreased in the area even though the manatees could still swim under the dock. Why did the population of manatees decrease?

A The sea grass grew too thick.

B The manatees died because they did not get enough light.

C The sea grass was poisoned.

D The sea grass died, and the manatees left the area.

2. A graph shows a direct linear relationship between the size of a fish and the amount of oxygen it uses. Which statement is true?

F The larger a fish is, the less oxygen the fish uses.

G The smaller the fish is, the less oxygen the fish uses.

H The smaller the fish is, the more oxygen the fish uses.

J It is impossible to determine a relationship.

3. The data below was collected during an experiment. What conclusion is supported by the data?

Temperature (°C)	Time to double bacteria population (min)
10	130
20	60
30	29
40	19
50	no growth

A Bacteria always grows faster at lower temperatures.

B The rate of growth of this bacterial species increases as temperature is increased from 10° C to 50° C.

C Temperature has no effect on the rate of bacteria growth.

D Bacteria do not like higher temperatures.

4. a. Why do scientists use graphs and charts during the evaluation of an experiment?

b. What types of graph would best show trends in data that change over time?

Copyright © Holt McDougal. All rights reserved.

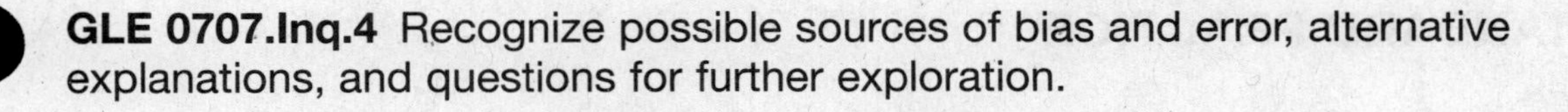

GLE 0707.Inq.4 Recognize possible sources of bias and error, alternative explanations, and questions for further exploration.

STANDARD REVIEW

Established theories are usually built on many experiments and observations. When new experimental results do not agree with the existing theory, scientists generally do not revise the theory immediately. Sometimes the results of the experiment contains an error in the data or the analysis. Even after an experiment has been reproduced independently, more information may be needed. Different scientists may interpret the results in different ways. In that case, further experiments are needed. The original experiment might suggest new ways to test the hypothesis.

Scientific investigations are a continual process. Even after results are reviewed and accepted by the scientific community for publication, the investigation of the topic may not be finished. New evidence may become available. The scientist may change the hypothesis based on the new evidence. In other cases, the scientist may have more questions that arise from the original evidence. When a number of different experiments all provide data that are not consistent with the theory, the theory is usually revised.

An example of a hypothesis that was not immediately accepted was Alfred Wegener's hypothesis of continental drift. Wegener analyzed fossils on several continents, and he found that some continents had fossils similar to fossils on other continents. After making many observations, Wegener proposed that the continents were once a supercontinent but later drifted apart. Although his results were confirmed by other scientists, Wegener's hypothesis was not generally accepted right away. Many scientists did not accept his hypothesis because it did not seem possible for the crust to move in the way he proposed. Until further investigations showed how the continents could move, continental drift was not generally accepted. Wegener's observations are now part of the support for the theory of *plate tectonics,* which describes the movement of the continents.

GUIDED PRACTICE

Directions: Using the Standard Review and what you have studied, read each question and circle the letter of the best response.

A report of which of the following observations would be inconsistent with current theories about Earth?

A Satellite data that show continents moving.

B Measurements show Earth's core is made of molten rock.

C Earth's revolution around the sun takes slightly more than 365 days.

D Observations show air pressure is lowest at sea level.

The correct answer is D. There is no current theory that would explain air pressure being lowest at sea level. The other observations are all consistent with our current understanding of how the planet functions.

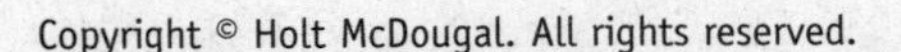

Copyright © Holt McDougal. All rights reserved.

STANDARD PRACTICE

1. Which of the following is <u>most likely</u> to result in a scientist drawing wrong conclusions?

A conducting a double-blind test

B conducting a controlled experiment

C gathering data for a brief period of time

D preventing opinions from affecting data collection

2. A scientist conducted an experiment to see how exposing turtle eggs to different conditions affects whether they produce male or female turtles. She organizes her results in the table below. Based on the data in the table, which of the following statements is true?

Light exposure	Temperature	Resulting turtles
sunlight	25.0°C	100% males
shade	28.5°C	50% males and 50% females
complete darkness	30.0°C	100% females

F A valid conclusion cannot be drawn because none of the variables changed.

G A valid conclusion cannot be drawn because too many of the variables changed.

H A valid conclusion is that as temperature increases, the number of males born increases.

J A valid conclusion is that as temperature increases, the number of females born increases.

3. If research groups studying the benefits of a new medicine report results that are quite different from each other, what can you conclude?

A One of the groups did not do the research correctly.

B You should assume both sets of results are not accurate.

C You should accept the results of the researchers at the larger medical center.

D More research is needed to determine what caused the results to be different.

4. a. Why does scientific knowledge change over time?

b. What should scientists do when new results are inconsistent with an existing, well-established theory?

Copyright © Holt McDougal. All rights reserved.

GLE 0707.Inq.5 Communicate scientific understanding using descriptions, explanations, and models.

STANDARD REVIEW

A key part of scientific research is communicating the results of an experiment. Scientific writing must present detailed information in a way that is very specific. That means it is very different from the literary writing that you use to tell a story or support a political point of view. Whether you are writing a laboratory report for your teacher or submitting a paper to a scientific journal, your lab report should contain enough information so that others can use it to reproduce your experiment and compare their results to yours.

A pattern, plan, representation, or description designed to show the structure or workings of an object, system, or concept is a *model.* With a model, a scientist can explain or analyze an object, system, or concept in more detail. Models are used in science to help explain how something works or to describe how something is structured.

After scientists have collected their data, they organize it into tables, charts, or graphs in order to look for trends and identify relationships among the variables. These visuals can show new information. For example, all of the data in a single row or column in a table have at least one characteristic in common. That characteristic is labeled in the row or column's label. In the table below, all of the values in the first column are temperatures given in degrees Celsius.

Temperature (°C)	Time to double bacteria population (min)
10	130
20	60
30	29
40	19
50	no growth

GUIDED PRACTICE

Directions: Using the Standard Review and what you have studied, read each question and circle the letter of the best response.

How does scientific writing differ from literary writing?

A The two types of writing have different purposes.

B Scientific reports are always written in English.

C Science writing is shorter than literary writing.

D Science writing is about facts, and literary writing is fiction.

The correct answer is A. Science writing must present details in a clear and detailed way, unlike literary writing. Scientific reports can be written in any language (Answer B), and they can be long or short, depending on the amount of information to be presented (Answer C). Nonfiction literary writing is also about facts (Answer D).

Copyright © Holt McDougal. All rights reserved.

STANDARD PRACTICE

1. Santos conducts an experiment. The data he collects does not support his hypothesis. What should Santos do when he writes a lab report for his experiment?

A He should report his data and observations accurately and honestly.

B He should change his data so they support his hypothesis.

C He should change the hypothesis to support the data.

D He should change both the hypothesis and the data.

2. The graph below shows population projections for the world at different growth rates. What does the graph project as the most likely outcome for the world population?

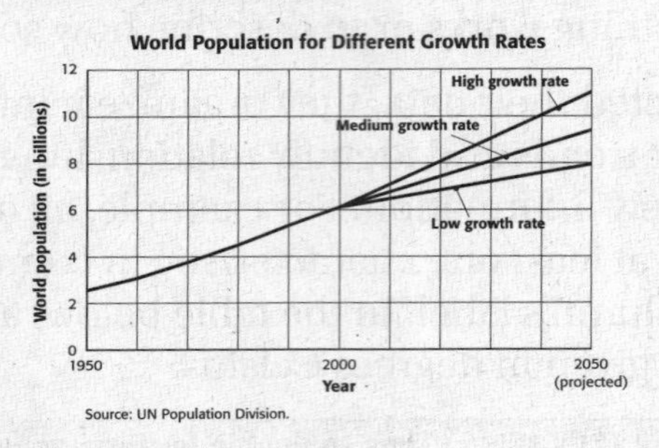

F It will be 12 billion in 2050.

G It will double by the year 2050.

H It will increase by 25% by the year 2050.

J It will be between 8 and 11 billion in 2050.

3. A representation of an object or system that helps scientists visualize and understand information is called a

A hypothesis.

B law.

C model.

D theory.

4. a. Why might two scientists working on the same problem draw different conclusions?

b. What should scientists do when there is more than one way to interpret a given set of findings?

Copyright © Holt McDougal. All rights reserved.

Grade 7 – Technology and Engineering

GLE 0707.T/E.1 Explore how technology responds to social, political, and economic needs.

STANDARD REVIEW

Scientists study the natural world. Engineers work to put scientific knowledge to practical use and to build the tools to use scientific knowledge. Some engineers design and build the buildings, roads, and bridges that make up cities. Others design and build electronic things, such as computers and televisions. Some even design processes and equipment to make chemicals and medicines. Engineers may work for universities, governments, and private companies.

While the driving force behind science is curiosity, the driving force behind technology is finding a solution to a social, political, or economic need. Technologies that societies need include power, communication, medical care, and transportation. Science and engineering together address these needs as they continually change. The internet, for example, combines many technologies to link distant parts of the world together in ways that were impossible even a few decades ago. Each communication advance leading to the internet combined scientific knowledge and engineering application. Political needs in a society include infrastructure used by everyone, such as roads and bridges. Another political need is defense, leading to the development of military technologies. Economic needs drive many technologies, including the development of better systems of manufacturing and distribution of materials.

Technology is a very broad term that includes many different types of application of science. It can be an object or device, such as a medical imaging machine. It can be a technique or method, such as a new way to make a fuel from grain crops. Technology can also be a system of production, for example, the assembly line, which changed how products were made throughout the world.

GUIDED PRACTICE

Directions: Using the Standard Review and what you have studied, read each question and circle the letter of the best response.

Which of the following is an example of technology?

A a pair of pliers

B a way to breed a new plant with desirable characteristics

C manufacturing robots that assemble automobiles

D all of the above

The correct answer is D. All of these are examples of different types of technology. A pair of pliers (Answer A) is not a newly discovered technology, but it is the application of scientific principles to make a tool. The technique described in Answer B is an example of a new method of doing something based on science. Robotic manufacturing (Answer C) is a technological system that has changed many manufacturing industries.

Copyright © Holt McDougal. All rights reserved.

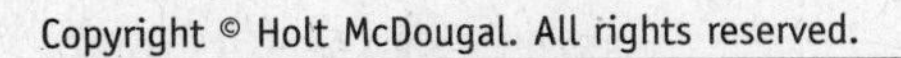

STANDARD PRACTICE

1. Which of these is research that had to be studied before optical fiber communication technologies could be developed?

A how light interacts with glass

B how computers work

C the number of people who travel into and out of a city each day

D how sound travels in a medium

2. Which type of technology was developed by scientists and engineers who studied biology and genetics?

F new types of ceramic glazes

G crops that are resistant to herbicides

H transmission of electrical energy

J fuel cells that run on hydrogen

3. Which type of technology would <u>most likely</u> be based on an understanding of Newton's laws?

A more efficient gasoline engine

B improved computer memory

C windows that provide better insulation

D treatments to prevent wood from rotting

4. a. How do technology advances to use alternative fuels in cars and buses benefit society?

b. What scientific research was needed in order to make electric vehicles that work reliably and economically?

Copyright © Holt McDougal. All rights reserved.

Grade 7 – Technology and Engineering

GLE 0707.T/E.2 Know that the engineering design process involves an ongoing series of events that incorporate design constraints, model building, testing, evaluating, modifying, and retesting.

STANDARD REVIEW

Technology is the process by which humans modify nature to meet their needs. Technology includes the products, equipment, and systems that you use every day, but it is more than that. Technology also includes the processes used to develop and build those products.

A major part of developing new technologies is engineering design. *Engineering* consists of the knowledge of the design to make products and develop processes to solve problems. As with scientists, the work of engineers must follow the laws of nature. Engineers also have to consider what materials are available, what safety problems may exist, and what the effects are on the environment.

In order to achieve their goals and develop new technologies, engineers use a design process. This process has many steps but it can be summarized as follows:

- identify the problem to be solved
- conduct research
- make decisions about materials and processes
- design and build models
- perform tests and evaluate the results
- modify the product or system
- test again and repeat the design process as many times as necessary
- build the technology

During this process, engineers may try many different materials and arrangements. When the change makes the product better, less expensive, more reliable or improved in any other way, it may be included in the next test. New scientific discoveries can become part of new design, making designs work even better.

GUIDED PRACTICE

Directions: Using the Standard Review and what you have studied, read each question and circle the letter of the best response.

As part of an engineering design process, engineers want to make a less expensive disposable towel. Which of these materials would they most likely want to test?

A silk　　**B** iron

C paper　　**D** fiberglass

The correct answer is C. Of the choices, paper is both the least expensive and most likely to work as a towel. Silk (Answer A) is a fabric that could be used, but it is much too expensive for a disposable product. Iron and fiberglass (Answers B and D) can both be woven into a flexible, cloth-like material, but it would probably not work as a towel. Both materials would also be too expensive for a disposable towel even if they could work for the purpose.

Copyright © Holt McDougal. All rights reserved.

STANDARD PRACTICE

1. Which of the following would be <u>most</u> important in designing a corn plant that produces a natural insecticide in its cells?

A cost of seed for the farmer

B height of the plant compared to other varieties

C not toxic to humans or domestic animals

D can be produced at a low cost

2. An engineer wants to design an instrument to measure the gases deep inside a volcano. Which of these would <u>not</u> be an important design problem?

F The instrument cannot be controlled remotely.

G The instrument housing corrodes quickly when exposed to acid.

H Data are not reliable when the temperature exceeds 50°C.

J There are only two companies that can build the instrument.

3. a. Why would an engineer designing a robot to study the surface of Mars need to build a prototype that operates on Earth before building the actual robot?

b. How could engineers be certain that the robot would work in the temperature and atmospheric condition it would find on Mars?

Copyright © Holt McDougal. All rights reserved.

Grade 7 – Technology and Engineering

GLE 0707.T/E.3 Compare the intended benefits with the unintended consequences of a new technology.

STANDARD REVIEW

The purpose of technology is to use science and engineering to solve problems and provide products and processes that people need. Technology has provided all the systems that hold our society together, including electric power, transportation networks, medical care, food supplies, and many others. Unfortunately, technology systems do not always work exactly as they are planned. Sometimes systems break down and stop working. Often, in addition to the planned benefits of technology, there are undesirable effects. These *unintended consequences* are problems caused by technology, which were not expected.

An example of an unintended consequence of technology is air pollution caused by burning fossil fuels. Transportation technology has certainly served its purpose. Automobiles, trucks, and airplanes provide the intended benefit, allowing people and goods to move easily from one place to another. At the same time, there have been unintended consequences of this technology. Acid rain caused by impurities in gasoline has damaged structures and destroyed stream ecosystems. Carbon dioxide released into the atmosphere when fuels burn is expected to cause dramatic changes in the world's climates. Another unintended consequence of transportation technology based on oil is political tension, which is the result of the uneven distribution of oil resources among different countries.

GUIDED PRACTICE

Directions: Using the Standard Review and what you have studied, read each question and circle the letter of the best response.

Which of these is an unintended consequence of technology?

A communicating with people in other countries by instant messages

B using your computer to find information about an illness from a library far from you home

C people finding your personal information by using their computers

D being able to purchase a CD online that is not available in a local store

The correct answer is C. One of the unintended consequences of the internet is that people can sometimes find information online that should not be public. The other choices are all intended benefits of internet technology: the ability to communicate easily (Answer A), find information quickly (Answer B) and shop at stores in other cities (Answer D).

Copyright © Holt McDougal. All rights reserved.

Grade 7 – Technology and Engineering

STANDARD PRACTICE

1. What could be an unintended consequence of altering the genes of honeybees so that they make more honey?

A increased production of honey
B bees that are more likely to attack their keepers
C bees that gather nectar more efficiently
D lower prices for honey at the grocery store

2. Which of the following is an example of an intended benefit of a new technology?

F A harder alloy allows drillers to reach oil far below Earth's surface.
G A tire that works better on wet roads does not last as long.
H A new clear plastic for windows turns cloudy when exposed to sunlight.
J A more energy-efficient car has a battery that must be replaced often.

3. Which of these effects is an unintended consequence of technology?

A A new traffic control design reduces travel time by one hour.
B A hydroelectric power plant damages the downstream ecosystems.
C A new plastic for car bodies has a finish that does not show scratches.
D A different type of elevator design allows buildings to be built taller.

4. a. What are some of the intended benefits of a dam built to produce hydroelectric power?

b. What are some possible unintended consequences of a dam built to produce hydroelectric power?

Copyright © Holt McDougal. All rights reserved.

Grade 7 – Technology and Engineering

GLE 0707.T/E.4 Describe and explain adaptive and assistive bioengineered products.

STANDARD REVIEW

Many technologies are based on new discoveries about life and how organisms change and adapt to their environments. *Bioengineering* applies the ideas of engineering to the science of living things. Chemical processes inside living cells control life and death, growth, reproduction, and how the cell interacts with its environment.

Scientists and engineers study the chemical reactions inside cells and learn how cells adapt to their environment. They also look at the structure of tissues and organs in plants and animals.

By using what they learn about living things, bioengineers develop new, useful products. Many of these products are used for health care. They use the knowledge of how cells operate to design new drugs that work better. Materials that can be implanted inside the human body are used to make artificial joints. Other new materials can replace body tissues, such as heart valves that are damaged by a viral infection. New bioengineered systems have allowed blind people to see and deaf people to hear by connecting devices directly to the body's nervous system. Bioengineers have also developed artificial arms and legs that can be controlled, through the nervous system, by a person's brain.

Bioengineers also work in other fields. In agriculture, for example, they modify the genes inside the cells of plants so that the plants kill harmful insects that eat their leaves. Other plants have been modified to produce bigger crop yields or to increase the amount of oil in their tissues, which can be extracted for use in fuel. These fuel oils from bioengineered plants are called *biofuels*.

In some cases, bioengineers can actually develop organisms to produce a material that is useful or performs a desirable function. Insulin, a medicine for treating diabetes, can be manufactured by bioengineered bacteria and then purified for medicinal use. Some bioengineered bacteria eat chemicals that are spilled in the environment. These bacteria have been modified so that they survive by eating oil or other chemicals absorbed by the ground or floating in water.

GUIDED PRACTICE

Directions: Using the Standard Review and what you have studied, read each question and circle the letter of the best response.

Which of these is the <u>best</u> example of a product developed by bioengineers?

A the internet

B cars that use less fuel

C bacteria that produce silk

D a new metal alloy

The correct answer is C. Although it is possible that bioengineering could have been used in the other products, the design of a new living cell is the best example of bioengineering because it modifies a living cell. The internet (Answer A) is based on electronics and system engineering, fuel-efficient cars (Answer B) are generally developed by mechanical engineering, and metal alloys are developed by chemical engineering.

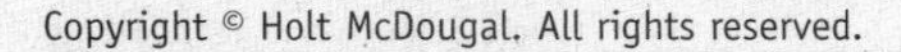

Copyright © Holt McDougal. All rights reserved.

Grade 7 – Technology and Engineering

STANDARD PRACTICE

1. How could the design of an artificial heart valve be advanced by bioengineering?

A by designing a polymer with the same characteristics as heart valve tissue

B by doing an internet search for companies that make heart valves

C by finding ways to replace the whole heart instead of just the valve

D by finding ways to prevent diseases that damage heart valves

2. Which of the following is <u>not</u> a reason manufacturers would use bioengineering in the development of new drugs?

F to find a less expensive way to make expensive drugs

G to increase the cost of manufacturing the drug

H to find a drug that is more closely related to natural products

J to improve the efficiency of a drug manufacturing process

3. a. Why could genetic engineering not be performed on cells before the discovery of the role of DNA in cells?

b. In what way is genetic engineering a form of technology?

Copyright © Holt McDougal. All rights reserved.

Grade 7 – Life Science: Standard 1 – Cells

GLE 0707.1.1 Make observations and describe the structure and function of organelles found in plant and animal cells.

STANDARD REVIEW

Plant cells have an outermost structure called a *cell wall,* a rigid structure that gives support to the cell. Plants and algae have cell walls made of a complex sugar called *cellulose.* All cells have a *cell membrane,* a protective barrier that encloses the cell. The cell membrane is made up of phospholipids that separate the cell's contents from its environment. The *cytoskeleton* is a web of proteins in the cytoplasm that acts as skeleton. It keeps the cell's membranes from collapsing.

Cells have organelles that carry out various life processes. *Organelles* are structures that perform specific functions within the cell. Different types of cells have different organelles. The table below summarizes the functions of different organelles.

Organelles and Their Functions

Nucleus the organelle that contains the cell's DNA and is the control center of the cell	**Chloroplast** the organelle that uses the energy of sunlight to make food
Ribosome the organelle in which amino acids are hooked together to make proteins	**Golgi complex** the organelle that processes and transports proteins and other materials out of cell
Endoplasmic reticulum the organelle that makes lipids, breaks down drugs and other substances, and packages proteins for Golgi complex	**Large central vacuole** the organelle that stores water and other materials
Mitochondrion the organelle that breaks down food molecules to make ATP	**Lysosome** the organelle that digests food particles, wastes, cell parts, and foreign invaders

GUIDED PRACTICE

Directions: Using the Standard Review and what you have studied, read each question and circle the letter of the best response.

Mitochondria are important organelles within a cell. What would <u>most likely</u> happen if a cell's mitochondria were <u>not</u> functioning properly?

A The cell's level of ATP would decrease.

B The cell's level of sugar would decrease.

C The cell would use lysosomes to release energy.

D The cell would create new mitochondria by cell division.

The correct answer is A. The mitochondria produce ATP, the source of energy for the cell. Sugar (Answer B) is produced by the chloroplasts of plant cells. Lysosomes do not release energy (Answer B) because their function is to process waste. Cell division reproduces the entire cell, not one particular organelle (Answer D).

Copyright © Holt McDougal. All rights reserved.

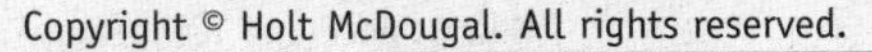

STANDARD PRACTICE

1. What is the name of a cell vesicle that contains enzymes that destroy damaged cell organelles?

A endoplasmic reticulum
B Golgi complex
C lysosome
D mitochondria

2. What is a cell membrane composed of?

F lipids
G nucleic acids
H phospholipids
J proteins

3. Where is most of the ATP in cells produced?

A endoplasmic reticulum
B ribosomes
C mitochondria
D Golgi complex

4. Photosynthesis occurs only in plant cells, but cellular respiration is a function of both plant and animal cells.

a. What cell structure is needed for photosynthesis to occur?

b. What cell structures are necessary for cellular respiration?

Copyright © Holt McDougal. All rights reserved.

Grade 7 – Life Science: Standard 1 – Cells

GLE 0707.1.2 Summarize how the different levels of organization are integrated within living systems.

STANDARD REVIEW

In a multicellular organism, groups of cells work together to perform specialized functions. A *tissue* is a group of cells that work together to perform a specific job. Animals have four basic types of tissue: nerve tissue, muscle tissue, connective tissue, and protective tissue. Plants have three types of tissue: transport tissue, protective tissue, and ground tissue. Transport tissue moves water and nutrients through a plant. Protective tissue covers the plant or animal and photosynthesis takes place in ground tissue.

A structure that is made up of two or more tissues working together to perform a specific function is called an *organ*. For example, your heart is an organ. A leaf is a plant organ that contains tissue that traps light energy to make food. A group of organs working together to perform a particular function is called an *organ system*. Each organ system has a specific job to do in the body.

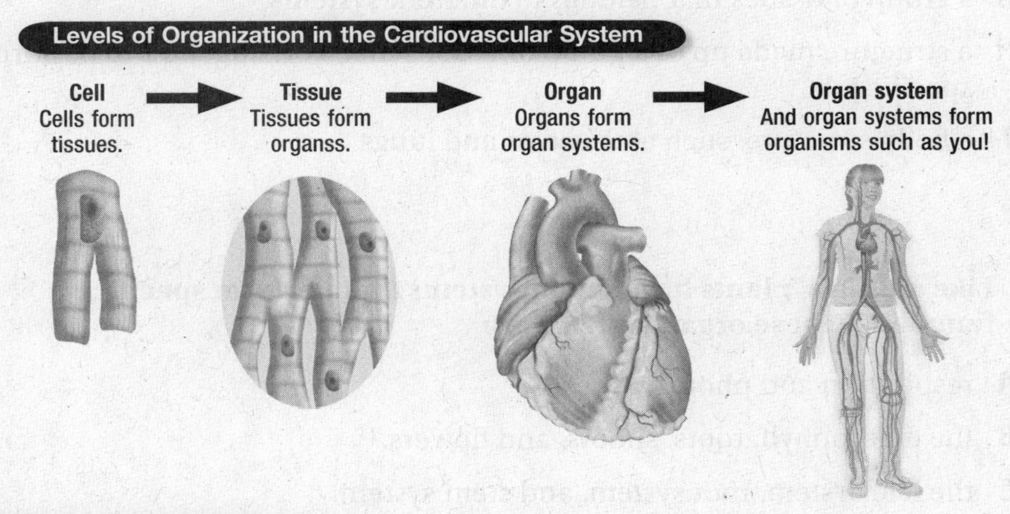

GUIDED PRACTICE

Directions: Using the Standard Review and what you have studied, read each question and circle the letter of the best response.

Which of the following is not a tissue found in the human body?

- **A** protective tissue
- **B** transport tissue
- **C** connective tissue
- **D** nerve tissue

The correct answer is B. All animal organisms, including humans, have protective, connective, and nerve tissues (Answers A, C and D). Transport tissues are found only in plants.

Copyright © Holt McDougal. All rights reserved.

Grade 7 – Life Science: Standard 1 – Cells

STANDARD PRACTICE

1. In a multicellular organism, each cell performs

A all functions.

B specific functions.

C random functions.

D the function it is trained for.

2. Which of the following best describes an organ?

F a group of cells that work together to perform a specific job

G a group of tissues that belongs to different systems

H a structure made up of a group of tissues that work together to perform a specific job

J a body structure, such as muscles and lungs

3. Like animals, plants have organ systems that perform specific functions. These organ systems are

A respiration and photosynthesis.

B the chlorophyll, roots, shoots, and flowers.

C the leaf system, root system, and stem system.

D the circulatory system and reproductive system.

4. a. Define the term "tissue" and explain how tissues fit within the organization of an organism.

b. Give an example of a type of tissue found in animals.

Copyright © Holt McDougal. All rights reserved.

GLE 0707.1.3 Describe the function of different organ systems and how collectively they enable complex multicellular organisms to survive.

STANDARD REVIEW

Your body's major organ systems all work together to perform the functions necessary for life. For example, the cardiovascular system, which includes the heart, blood, and blood vessels, works with the respiratory system, which includes the lungs. The cardiovascular system picks up oxygen from the lungs and carries the oxygen to cells in the body. These cells produce carbon dioxide, which the cardiovascular system returns to the respiratory system. The respiratory system expels the carbon dioxide.

Organ System	Function
Integumentary System	Skin, hair, and nails protect other tissues.
Muscular System	Works with the skeletal system to help you move.
Skeletal System	Provides a frame to support and protect body parts.
Cardiovascular System	Distributes blood through blood vessels to other organs.
Respiratory System	Brings in oxygen and releases carbon dioxide.
Urinary System	Removes wastes from the blood and regulates body fluids.
Reproductive System	The female reproductive system produces eggs and nourishes and protects the fetus. The male reproductive system produces and delivers sperm.
Nervous System	Receives and sends electrical messages throughout your body.
Digestive System	Breaks down the food into nutrients that the body can use.
Lymphatic System	Returns fluids to blood vessels and helps get rid of bacteria and viruses.
Endocrine System	Sends out chemical messages that control other systems.

GUIDED PRACTICE

Directions: Using the Standard Review and what you have studied, read each question and circle the letter of the best response.

Which body system includes the lungs?

- **A** cardiovascular
- **B** respiratory
- **C** muscular
- **D** skeletal

The correct answer is B. The lungs add oxygen and remove carbon dioxide from the blood as one of the functions of the respiratory system. The cardiovascular system (Answer A) pumps the blood through the body; the muscular system (Answer C) provides motion to parts of the body; and the skeletal system (Answer D) provides the framework that supports the rest of the body.

Copyright © Holt McDougal. All rights reserved.

STANDARD PRACTICE

1. The path of food through the digestive tract is

A mouth, esophagus, stomach, liver, small intestine and large intestine.

B mouth, esophagus, stomach, small intestine and large intestine.

C mouth, esophagus, stomach, gallbladder, small intestine and large intestine.

D mouth, stomach, esophagus, small intestine and large intestine.

2. What body system is shown in the illustration?

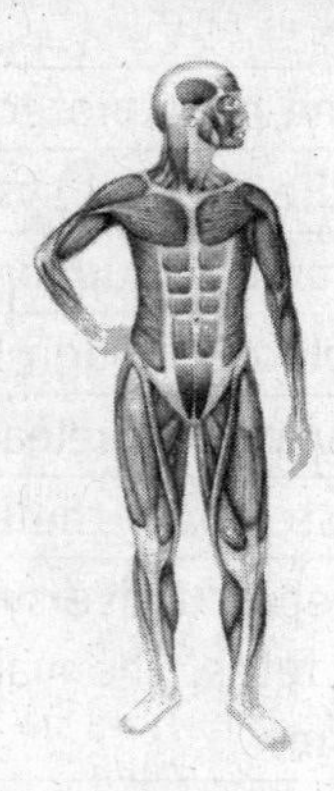

F skeletal system

G integumentary system

H muscular system

J cardiovascular system

3. Which system includes your toenails?

A skeletal system

B integumentary system

C muscular system

D cardiovascular system

4. a. Which two body systems play a major role in delivery of oxygen throughout your body?

b. Which of the systems in the table shown on the previous page, are necessary for a person to remain alive? Explain your answer.

Copyright © Holt McDougal. All rights reserved.

Grade 7 – Life Science: Standard 1 – Cells

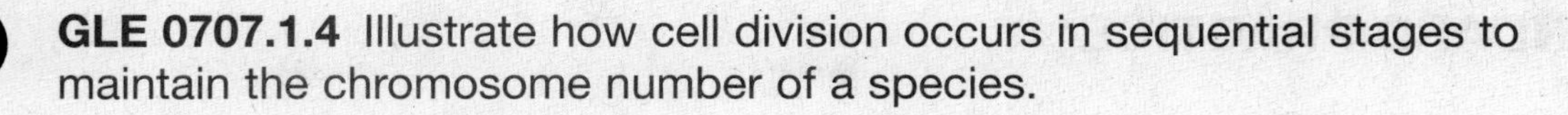

GLE 0707.1.4 Illustrate how cell division occurs in sequential stages to maintain the chromosome number of a species.

STANDARD REVIEW

The life cycle of a cell is called the *cell cycle.* The cell cycle begins when the cell is formed and ends when the cell divides and forms new cells. Before a cell divides, it must make a copy of its *deoxyribonucleic acid (DNA),* the hereditary material that controls all cell activities, including the making of new cells. The DNA of a cell is organized into structures called *chromosomes.* Copying chromosomes ensures that each new cell will be an exact copy of its parent cell. The copied chromosomes separate, and the cell splits into two new, identical cells, in a process called *mitosis.*

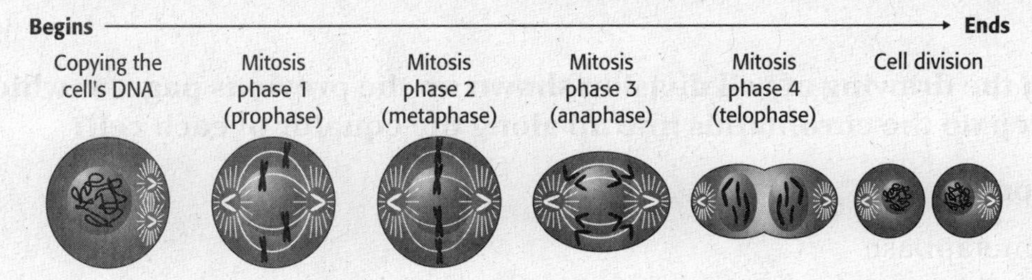

Interphase - Before mitosis begins, chromosomes are copied. Each chromosome then exists as two chromatids.
Mitosis Phase 1 (Prophase) The nuclear membrane dissolves, and chromosomes condense into rod like structures.
Mitosis Phase 2 (Metaphase) Chromosomes line up in pairs along the center of the cell.
Mitosis Phase 3 (Anaphase) Chromatids separate and move to opposite sides of the cell.
Mitosis Phase 4 (Telophase) A nuclear membrane forms around each set of chromosomes.
Cytokinesis In cells that lack a cell wall, the cell pinches in two. In cells that have a cell wall, a cell plate forms between the two new cells.

GUIDED PRACTICE

Directions: Using the Standard Review and what you have studied, read each question and circle the letter of the best response.

What is the end result of mitosis?

A two identical cells
B two cells that are related but different
C two cell nuclei that will become cells
D one cell with twice as many chromosomes

The correct answer is A. Mitosis builds two identical cells by making two copies of the DNA, which tells the cell how to grow, so Answer B is not correct because the cells are identical. Answer C is incorrect because mitosis produces two complete cells. Answer D describes the beginning phase of mitosis, not the end.

Copyright © Holt McDougal. All rights reserved.

STANDARD PRACTICE

1. Which of these cells would form a cell plate during mitosis?

A animal cells

B plant cells

C human cells

D all of the above

2. In the drawing of cell division shown on the previous page, at which step do the chromatids line up along the equator of each cell?

F prophase

G metaphase

H anaphase

J telophase

3. Once mitosis is completed, the process of cytokinesis occurs. What happens during this process in animal cells?

A The cytoplasm splits in two.

B The nuclear membrane breaks apart.

C The chromosomes line up.

D The chromatids separate.

4. a. How does mitosis ensure that a new cell is just like its parent cell?

b. What happens to the cytoplasm and other materials in the cell when it divides?

Copyright © Holt McDougal. All rights reserved.

GLE 0707.1.5 Observe and explain how materials move through simple diffusion.

STANDARD REVIEW

Water is made up of molecules, which are in constant motion. Molecules travel from where they are crowded to where they are less crowded. This movement from areas of high concentration (crowded) to areas of low concentration (less crowded) is called *diffusion.* Pure water has the highest concentration of water molecules. When you mix something, such as food coloring, sugar, or salt, with water, you lower the concentration of water molecules. The membrane of a cell is *semipermeable,* which means that only certain substances can pass through. Water molecules can move through the membrane from liquid with a high concentration of water molecules to the liquid with the lower concentration of water molecules. However, other substances dissolved in the water, such as sugars and salts cannot pass through the membrane.

The cells of organisms are surrounded by and filled with fluids that are made mostly of water. The diffusion of water through cell membranes is called *osmosis.* Osmosis is important to cell functions. For example, red blood cells are surrounded by plasma. Plasma is made up of water, salts, sugars, and other particles. The concentration of these particles is kept in balance by osmosis. If red blood cells were in pure water, water molecules would flood into the cells and cause them to burst. When red blood cells are put into a salty solution, the concentration of water molecules inside the cell is higher than the concentration of water outside. This difference makes water move out of the cells, and the cells shrivel up. Osmosis also occurs in plant cells. When a wilted plant is watered, osmosis makes the plant firm again.

GUIDED PRACTICE

Directions: Using the Standard Review and what you have studied, read each question and circle the letter of the best response.

What happens to red blood cells when they are surrounded by pure water?

- **A** The cells swell as water enters by osmosis.
- **B** The cells shrink as water diffuses out of them.
- **C** The cell membrane blocks water from moving in or out of the cell.
- **D** The cells lose salt and sugar by osmosis.

The correct answer is A. Osmosis causes the water to move from where it is more concentrated to where it is less concentrated. Cells do not shrink (Answer B) because they do not lose water. The cell membrane allows water to pass through (Answer C) but not salts or sugars (Answer D).

Copyright © Holt McDougal. All rights reserved.

Grade 7 – Life Science: Standard 1 – Cells

STANDARD PRACTICE

1. The figure below illustrates the process of osmosis. Why do the levels of liquids on the two sides of the membrane change?

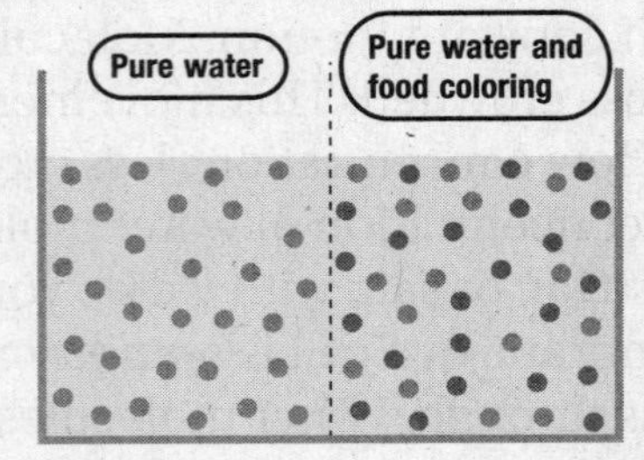

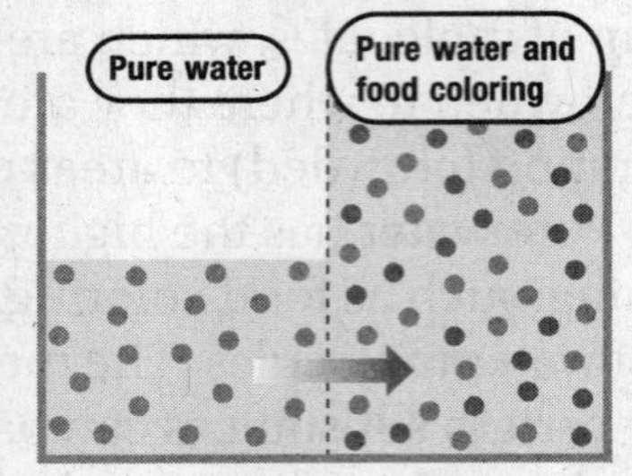

A Water flows to the side with the lower food coloring concentration.

B Food coloring flows to the side with the lower food coloring concentration.

C Water and food coloring flow back and forth to make the concentrations the same.

D Water diffuses to the area with the lower water concentration.

2. What happens to blood cells if the concentration of salt outside the cells is higher than the concentration of salt inside the cell?

F Water diffuses into the cell.

G Water diffuses out of the cell.

H The cell expands and ruptures.

J The cell manufactures salt to make the concentrations equal.

3. Which of these compounds can pass through a cell membrane?

A water

B salt

C sugar

D all of the above

4. a. How does your body control the concentration of water inside cells?

b. Why is it important that the concentrations of salt and sugar in the blood be maintained at the correct levels?

Copyright © Holt McDougal. All rights reserved.

Grade 7 – Life Science: Standard 3 – Flow of Matter and Energy

GLE 0707.3.1 Distinguish between the basic features of photosynthesis and respiration.

STANDARD REVIEW

Photosynthesis is the process by which cells, such as plant cells, use sunlight, carbon dioxide, and water to make sugar and oxygen. Photosynthesis takes place in a cell's *chloroplasts.* Chloroplasts are green because they contain *chlorophyll,* a green pigment that is found in an internal membrane system that traps the energy of sunlight. This energy is used to make sugar. The sugar produced by photosynthesis is then used by mitochondria to make ATP.

Photosynthesis

$$6CO_2 + 6H_2O \rightarrow C_6H_{12}O_6 + 6H_2O$$

Carbon dioxide + Water → Glucose + Oxygen

Sugars made by chloroplasts are used for energy by the plants and by animals that eat the plants. Sugars are processed in the mitochondria through cellular respiration. *Cellular respiration* uses oxygen to break down food for energy. A *mitochondrion* is the main power source of a cell. Most eukaryotic cells have mitochondria. Energy released by mitochondria is stored in a substance called ATP. The cell then uses ATP to do work. Most of a cell's ATP is made on the inner membrane of the cell's mitochondria. The chemical reaction for cellular respiration is shown below.

Cellular Respiration

$$C_6H_{12}O_6 + 6H_2O \rightarrow 6CO_2 + 6H_2O$$

Glucose + Oxygen → Carbon dioxide + Water

GUIDED PRACTICE

Directions: Using the Standard Review and what you have studied, read each question and circle the letter of the best response.

What is the source of energy for photosynthesis in plant cells?

A ATP and other chemicals

B carbon dioxide

C chlorophyll

D sunlight

The correct answer is D. Plants use the energy from sunlight to make sugars, using the materials carbon dioxide (Answer B) and water. Chlorophyll in the plant absorbs the energy of sunlight (Answer C). ATP is one of the chemicals that plant and animal cells use to transfer energy to provide power for cell processes (Answer A).

Copyright © Holt McDougal. All rights reserved.

STANDARD PRACTICE

1. Which of these is the organelle responsible for cellular respiration in both animals and plants?

A chloroplast

B nucleus

C mitochondrion

D ATP

2. What are the products of cellular respiration?

F carbon dioxide and oxygen

G glucose and water

H carbon dioxide and water

J glucose and ATP

3. Which of the following body responses is a sign that your cells need more energy?

A Your breathing rate increases.

B You begin to shiver.

C You feel hungry.

D You feel thirsty.

4. a. Why are plants important to the survival of all living things?

b. What happens to the solar energy absorbed by chlorophyll?

Copyright © Holt McDougal. All rights reserved.

Grade 7 – Life Science: Standard 3 – Flow of Matter and Energy

GLE 0707.3.2 Investigate the exchange of oxygen and carbon dioxide between living things and the environment.

STANDARD REVIEW

The early atmosphere of Earth did not contain oxygen gas. There is evidence that cyanobacteria first appeared more than 3 billion years ago and began to use sunlight to produce their own food. This process releases oxygen. The first cyanobacteria began to release oxygen gas into the oceans and air and it eventually reached the current level, about 20% of the air.

Today, plants and other photosynthetic organisms, such as some bacteria and many protists, form the base of nearly all food chains on Earth. In addition to relying on them for food, almost all organisms need the oxygen that they produce. Plants, animals, and most other organisms rely on cellular respiration to get energy. Cellular respiration requires oxygen, which is a byproduct of photosynthesis. So, photosynthesis provides the oxygen that animals and plants need for cellular respiration. Cellular respiration, in turn, provides the carbon dioxide needed for photosynthesis to occur. Photosynthesis cannot occur without carbon dioxide. Oxygen and carbon dioxide are continually produced in a cyclic process.

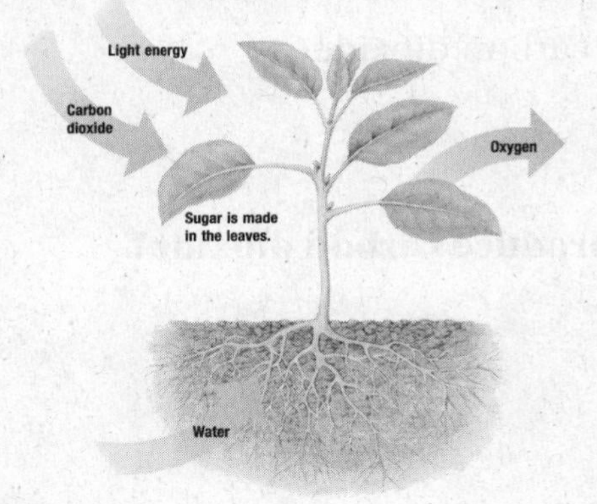

GUIDED PRACTICE

Directions: Using the Standard Review and what you have studied, read each question and circle the letter of the best response.

What is the source of most of the oxygen in the atmosphere?

A cellular respiration

B photosynthesis

C sunlight

D the ocean

The correct answer is B. Photosynthesis is the process by which plants use carbon dioxide and water to produce sugar and oxygen. Cellular respiration (Answer A), the opposite process, consumes oxygen. Sunlight (Answer C) provides the energy for photosynthesis, but it does not produce oxygen. The oceans (Answer D) do not produce oxygen.

Copyright © Holt McDougal. All rights reserved.

Grade 7 – Life Science:
Standard 3 – Flow of Matter and Energy

STANDARD PRACTICE

1. What happens during photosynthesis in green plants?

A Plants produce food.

B Plants produce oxygen.

C Plants consume carbon dioxide.

D all of the above

2. What gas is necessary for cellular respiration to occur?

F oxygen

G carbon dioxide

H nitrogen

J both oxygen and carbon dioxide

3. What organisms produce carbon dioxide?

A plants

B animals

C bacteria

D all of the above

4. a. Why could plants not have existed before cyanobacteria started producing oxygen?

b. What is the primary source of carbon dioxide in the atmosphere today?

Copyright © Holt McDougal. All rights reserved.

Grade 7 – Life Science: Standard 4 – Heredity

GLE 0707.4.1 Compare and contrast the fundamental features of sexual and asexual reproduction.

STANDARD REVIEW

There are two kinds of reproduction: asexual and sexual. *Asexual reproduction* results in offspring with genotypes that are exact copies of their parent's genotype. *Sexual reproduction* produces offspring that share traits with their parents, but are not exactly like either parent.

In sexual reproduction, two parent cells join together to form offspring that are different from both parents. (Figure A) The parent cells are called *sex cells.* Sex cells are different from ordinary body cells. Every organism has a characteristic number of chromosomes. Human body cells have 46 chromosomes, organized in 23 pairs. Human sex cells are different, having only 23 chromosomes. Sex cells are made during meiosis. *Meiosis* is a copying process that produces cells with half the usual number of chromosomes. For example, a human egg cell has 23 chromosomes, and a sperm cell has 23 chromosomes. The new cell that forms when an egg cell and a sperm cell join has 46 chromosomes.

In asexual reproduction, only one parent cell is needed. (Figure B) The structures inside the cell are copied, and then the parent cell divides, making two exact copies. This type of cell reproduction is known as *mitosis.* Most of the cells in your body and most single-celled organisms reproduce in this way.

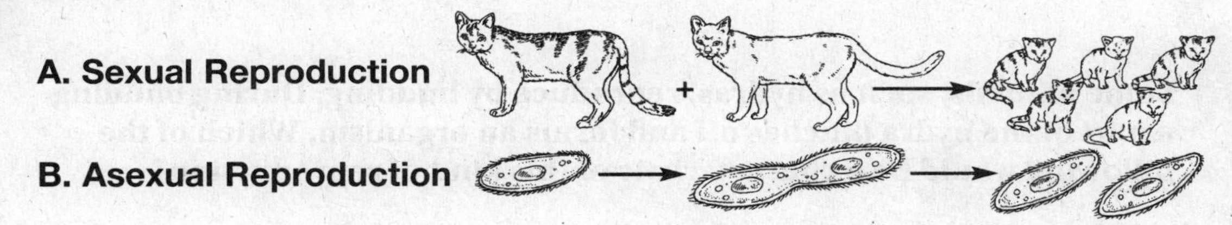

GUIDED PRACTICE

Directions: Using the Standard Review and what you have studied, read each question and circle the letter of the best response.

When compared with the original cell, the cells produced during meiosis

A are identical to the original cell.

B have half the number of chromosomes.

C have double the number of chromosomes.

D lack any chromosomes.

The correct answer is B. Meiosis is the process by which sex cells are produced for sexual reproduction. Each sex cell has half the normal number of chromosomes. Identical cells are formed by mitosis (Answer A). There are no cell division processes that result in cells with twice as many chromosomes (Answer C) or no chromosomes at all (Answer D).

Copyright © Holt McDougal. All rights reserved.

STANDARD PRACTICE

1. **Most of the cells in your body reproduce**

A by meiosis.

B asexually.

C sexually.

D by joining cells.

2. **How many chromosomes occur in the new cell formed when human male and female sex cells carrying genetic information unite?**

F 23

G 32

H 46

J 92

3. **Some animals, such as hydras, reproduce by budding. During budding, a part of the hydra pinches off and forms an organism. Which of the following would be a characteristic of this kind of reproduction?**

A Genetic variation occurs among offspring.

B Two parents are required to contribute genetic information.

C Offspring take a long time to develop.

D Offspring are genetically identical to the parent.

4. **a. Explain the difference between the offspring of sexual and asexual reproduction.**

b. Give an example of each type of reproduction.

Copyright © Holt McDougal. All rights reserved.

Grade 7 – Life Science: Standard 4 – Heredity

GLE 0707.4.2 Demonstrate an understanding of sexual reproduction in flowering plants.

STANDARD REVIEW

In flowering plants, fertilization takes place within flowers. *Pollination* happens when pollen is moved from anthers to stigmas. Usually, wind or animals move pollen from one flower to another flower. Pollen contains sperm. After pollen lands on the stigma, a tube grows from each pollen grain. The tube grows through the style to an ovule. Ovules are found inside the ovary. Each ovule contains an egg. Sperm from the pollen grain moves down the pollen tube and into an ovule. *Fertilization* occurs when a sperm fuses with an egg inside an ovule.

After fertilization, the ovule develops into a seed. The seed contains a tiny, undeveloped plant. The ovary surrounding the ovule becomes a fruit. As a fruit swells and ripens, it protects the developing seeds. Once a seed is fully developed, the young plant inside the seed stops growing. When seeds are dropped or planted in a suitable environment, the seeds sprout. To sprout, most seeds need water, air, and warm temperatures.

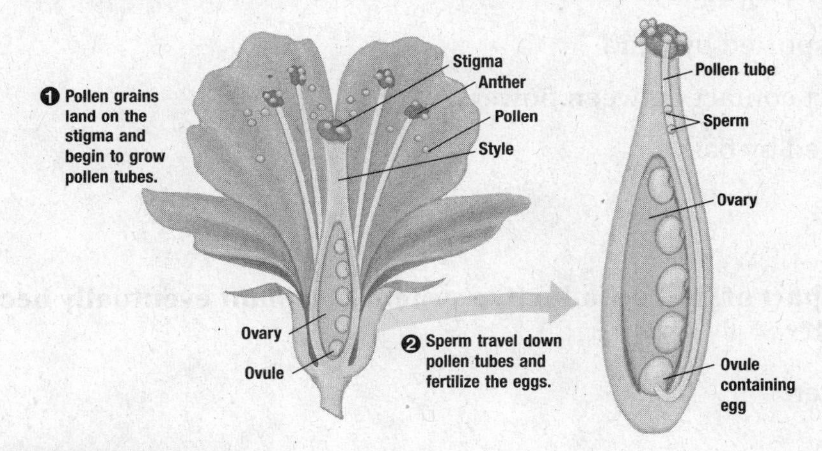

GUIDED PRACTICE

Directions: Using the Standard Review and what you have studied, read each question and circle the letter of the best response.

In the sexual reproduction of flowering plants, fertilization occurs when

A pollen is carried from one flower to another.

B pollen is deposited on the stigma.

C the sperm and egg cells combine.

D an undeveloped plant forms in the ovule.

The correct answer is C. Fertilization is the combination of sperm and egg to form a new cell. Prior to fertilization, a plant produces pollen, which is carried to the stigma of the same plant or another plant (Answers A and B). After fertilization, a seed, which contains an undeveloped plant, forms (Answer D).

Copyright © Holt McDougal. All rights reserved.

STANDARD PRACTICE

1. What part of the reproductive system of a plant eventually becomes a fruit?

A egg cell

B ovary

C ovule

D stigma

2. Which of these is not a method by which pollen is transferred from one plant to another?

F carried by honeybees

G transported by wind

H direct contact between flowers

J carried by bats

3. What part of the reproductive system of a plant eventually becomes a seed?

A anther

B ovary

C ovule

D style

4. a. What advantage do plants gain by having seeds remain dormant?

b. What do dormant seeds need in order to germinate?

Copyright © Holt McDougal. All rights reserved.

Grade 7 – Life Science: Standard 4 – Heredity

GLE 0707.4.3 Explain the relationship among genes, chromosomes, and inherited traits.

STANDARD REVIEW

Organisms use coded information to pass genetic information from one generation to the next. The instructions for inherited characteristics are carried by genes. A *gene* is a segment of **d**eoxyribo**n**ucleic **a**cid (DNA) that contains instructions for making a protein or signaling a cell to perform some function. Variations in genes from one organism to another account for the broad range of characteristics of living things.

Genes are located in structures called *chromosomes,* which are in the nucleus of a cell, or in the cytoplasm of organisms that do not have nuclei. A chromosome is made up of proteins and a single large molecule of DNA. The DNA molecules in the chromosomes include all the genes that carry instructions for the organism's traits.

In asexual reproduction, each chromosome is copied, and one copy is passed to each of the two new cells. Offspring of sexual reproduction receive half their genes from each parent, so the new cell is not a duplicate of either parent. Each parent gives one set of genes to the offspring. The offspring then has two forms of the same gene for every characteristic—one from each parent. The different forms of a gene are known as *alleles.* In some cases, only one of the alleles determines a characteristic of an organism, although the other allele can be passed to the next generation. In other cases, both alleles affect the trait. As a result, there is a greater variation of inherited characteristics among individuals of species that reproduce sexually.

GUIDED PRACTICE

Directions: Using the Standard Review and what you have studied, read each question and circle the letter of the best response.

Which of the following is <u>not</u> involved in passing inherited characteristics from one generation to the next?

A deoxyribonucleic acid

B cytoplasm

C chromosomes

D genes

The correct answer is B. Cytoplasm makes up most of the volume of the cell contents, but it is not directly involved in reproduction. Genetic information is recorded on genes (Answer D), which are composed of deoxyribonucleic acid (Answer A). Inside the cell, genes are located on the chromosomes (Answer C).

Copyright © Holt McDougal. All rights reserved.

STANDARD PRACTICE

1. Genes carry information that determines

A alleles.
B ribosomes.
C chromosomes.
D traits.

2. Why is there more genetic variation in organisms that reproduce sexually than in organisms that reproduce asexually?

F Organisms that reproduce sexually have more cells.
G Organisms that reproduce sexually receive two different sets of genes.
H Organisms that reproduce have more chromosomes.
J Organisms that reproduce asexually are smaller.

3. The passing of traits from parents to offspring is called

A probability.
B heredity.
C recessive.
D meiosis.

4. a. Where are the genes located inside a cell?

b. What is the function of a gene within the cell?

Copyright © Holt McDougal. All rights reserved.

Grade 7 – Life Science: Standard 4 – Heredity

GLE 0707.4.4 Predict the probable appearance of offspring based on the genetic characteristics of the parents.

STANDARD REVIEW

Genetics is the study of how traits are inherited. *Genes* carry the information that determines the traits of offspring. An organism's appearance, which is based on these traits, is known as its *phenotype.* In a famous experiment with pea plants, Gregor Mendel studied the inheritance of either purple flowers or white flowers. The phenotypes for flower color among these plants are purple and white. Mendel determined that there are two sets of instructions for each inherited characteristic. Each parent gives one set of genes to the offspring. The offspring then has two forms of the same gene for every characteristic—one from each parent. The different types of a single gene are called *alleles.* In many cases, one allele is dominant and the other is recessive. The organism exhibits the dominant trait, but it can pass the allele for the recessive trait to its offspring without the trait being expressed. In the pea plant experiment, the dominant allele was for purple flowers. Pea plants with one or both purple flower alleles had purple flowers. Only plants with two of the recessive white flower alleles had white flowers.

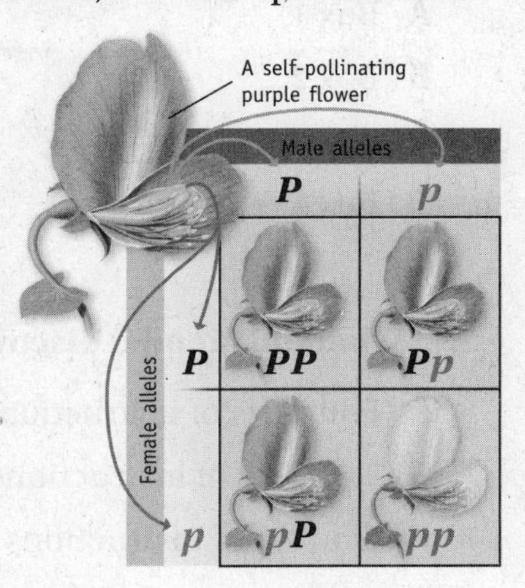

A *Punnett square,* like the one shown, is used to organize the possible genotypes that offspring can inherit from a particular pair of parents and predict the traits of the offspring. Dominant traits are indicated by a capital letter and recessive traits by a small case letter.

GUIDED PRACTICE

Directions: Using the Standard Review and what you have studied, read each question and circle the letter of the best response.

A white flower that has the genotype *pp* is crossed with a purple flower that has the genotype *PP*. What are the possible genotypes of the offspring?

A *PP* and *pp*

B *PP*

C *Pp*

D *pp*

The correct answer is C. Because each parent has only one allele, it can only contribute that allele. That means that all offspring must inherit one of each. Answers A, B and D only include genotypes with identical alleles, so they are incorrect.

Copyright © Holt McDougal. All rights reserved.

Grade 7 – Life Science: Standard 4 – Heredity

STANDARD PRACTICE

1. In humans, having freckles (F) is dominant, and having freckles is recessive (f). In the Punnett square below, which box contains the gene combination that will not produce freckles?

A Box 1

B Box 2

C Box 3

D Box 4

2. What did Mendel discover in his studies of pea plant reproduction?

F Four sets of instructions must be present for each characteristic.

G Two sets of instructions must be present for each characteristic.

H One set of instructions must be present for each characteristic.

J No sets of instructions must be present for each characteristic.

3. The instructions that determine an offspring's inherited traits are called

A alleles.

B phenotype.

C chromosomes.

D genes.

4. a. Define the terms gene and allele.

b. What determines an organism's alleles in sexual reproduction?

Copyright © Holt McDougal. All rights reserved.

Grade 7 – Earth and Space Science: Standard 7 – The Earth

GLE 0707.7.1 Describe the physical properties of minerals.

STANDARD REVIEW

You may think that all minerals look like gems, but in fact, most minerals look more like rocks. A *mineral* is a naturally formed, inorganic solid that has a definite crystalline structure. Solid, geometric forms of minerals produced by a repeating pattern of atoms or molecules that is present throughout the mineral are called *crystals.* A crystal's shape is determined by the arrangement of the atoms or molecules within the crystal. The arrangement of atoms or molecules in turn is determined by the kinds of atoms or molecules that make up the mineral.

Minerals are classified by a number of physical properties. The way a surface reflects light is called *luster.* If a mineral is shiny, it has a metallic luster. If the mineral is dull, its luster is either submetallic or nonmetallic. The color of a mineral in powdered form is called the mineral's *streak.* A mineral's streak can be found by rubbing the mineral against a piece of unglazed porcelain called a *streak plate.* The streak is a thin layer of powdered mineral. Different types of minerals break in different ways. The way a mineral breaks is determined by the arrangement of its atoms. *Cleavage* is the tendency of some minerals to break along smooth, flat surfaces. *Fracture* is the tendency of some minerals to break unevenly along curved or irregular surfaces. A mineral's resistance to being scratched is called *hardness.* To determine the hardness of minerals, scientists use the *Mohs hardness scale.* Talc, which is very soft, has a rating of 1, and diamond, which is very hard, has a rating of 10. The greater a mineral's resistance to being scratched, the higher the mineral's rating. *Density* is the measure of how much matter is in a given amount of space. Each mineral has a characteristic density based on how its atoms are packed together. The same mineral can come in a variety of colors, depending on impurities. Although color is sometime useful, it is generally not the best way to identify a mineral.

GUIDED PRACTICE

Directions: Using the Standard Review and what you have studied, read each question and circle the letter of the best response.

Which property of minerals does the Mohs scale measure?

A luster

B hardness

C density

D streak

The correct answer is B. Minerals are rated on the numerical Mohs scale based on their resistance to scratching. A mineral higher on the scale will scratch a mineral lower on the scale. Luster (Answer A) is determined by appearance. Density (Answer C) is determined by measuring the amount of mass in a given volume. Streak (Answer D) is the color and appearance of the powdered mineral.

Copyright © Holt McDougal. All rights reserved.

STANDARD PRACTICE

1. Why does the streak of a mineral often have a different color than a large piece of the mineral?

A The streak has not been exposed to oxygen in the air.

B The mineral reacts with the streak plate.

C The mineral composition changes when it is broken apart.

D The streak is composed of larger particles.

2. Which of these minerals has a metallic luster?

F gypsum

G quartz

H galena

J calcite

3. Which of these minerals could be used to scratch corundum, which is rated 9 on the Mohs hardness scale.

A talc

B diamond

C quartz

D topaz

4. a. Why is ice considered a mineral?

b. How is a mineral different from a rock?

Copyright © Holt McDougal. All rights reserved.

Grade 7 – Earth and Space Science: Standard 7 – The Earth

GLE 0707.7.2 Summarize the basic events that occur during the rock cycle.

STANDARD REVIEW

Geologists place all rocks into three major classes – sedimentary, igneous, and metamorphic – based on how the rocks form. *Sedimentary rocks* form when existing rocks break into smaller pieces, and those pieces become cemented together. *Igneous rocks* form when hot, molten rock, called *magma,* cools and becomes solid. *Metamorphic rocks* form when other rock is changed by chemical processes, temperature, or pressure. The processes in which a rock forms, changes from one type to another, is destroyed, and forms again by geological processes is called the *rock cycle.* In the cycle, water, wind, ice, and heat break rock into fragments. These rock and mineral fragments, or *sediment,* are eventually buried, then compacted and cemented together to form sedimentary rock. When large masses of Earth's crust collide, some of the rock is forced downward. Beneath the surface, intense heat and pressure change the sedimentary rock into metamorphic rock. The hot liquid that forms when rock partially or completely melts is called *magma.* When magma rises toward the surface, it cools and solidifies to become igneous rock.

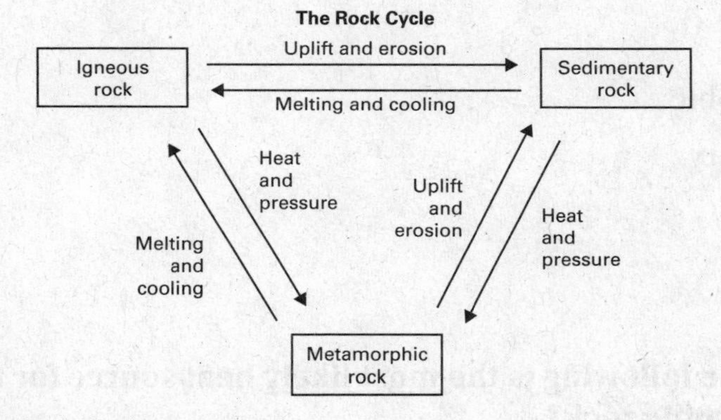

GUIDED PRACTICE

Directions: Using the Standard Review and what you have studied, read each question and circle the letter of the best response.

Analyze the diagram to determine which of the following steps in the rock cycle is <u>necessary</u> for sedimentary rock to form.

- **A** heat and pressure
- **B** melting and cooling
- **C** uplift and erosion
- **D** volcanic activity

The correct answer is C. Sediment forms when rock is lifted to the surface and then erodes. Heat and pressure (Answer A) are involved in the conversion to metamorphic rock, while melting and then cooling (Answer B) is the process that forms igneous rock. Volcanic activity (Answer D) is part of this cooling process.

Copyright © Holt McDougal. All rights reserved.

Grade 7 – Earth and Space Science: Standard 7 – The Earth

STANDARD PRACTICE

1. Which of the following statements is the best summary of the rock cycle?

A Rocks deep below ground rise to the surface, are moved back underground, then rise to the surface again.

B Igneous rock and sedimentary rock change to metamorphic rock.

C The rock cycle has a single pathway from one type of rock to another type of rock.

D Every type of rock can be changed into every other type of rock. The type of rock that forms depends on the conditions that affect the rock.

2. What type of rock forms when heat and pressure change the structure, texture, or composition of sedimentary rock?

F igneous

G metamorphic

H sedimentary

J clastic

3. Which of the following is the most likely heat source for the formation of metamorphic rock?

A the heat from inside Earth

B solar energy

C uplift by colliding plates

D the friction of movement of the plates against one another

4. a. Describe the conditions and environments under which igneous rock and metamorphic rock form.

b. Describe the conditions and environments under which sedimentary rock forms and how these processes fit into the rock cycle as a whole.

Copyright © Holt McDougal. All rights reserved.

Grade 7 – Earth and Space Science: Standard 7 – The Earth

GLE 0707.7.3 Analyze the characteristics of the earth's layers and the location of the major plates.

STANDARD REVIEW

The outermost layer of Earth is the *crust.* The crust is 5 to 100 km thick. It is the thinnest layer of Earth. There are two types of crust—continental and oceanic. This crust forms a large plate that floats on the material beneath it, constantly moving and reshaping the surface of Earth.

The layer of Earth beneath the crust is the *mantle.* The rock of the mantle is so hot that it flows like thick syrup, pushing the plates that rest on it. The mantle is much thicker than the crust and contains most of Earth's mass. In some places, mantle rock pushes to the surface, which allows scientists to study the rock directly.

Beneath the mantle, extending to the center of the planet, is Earth's *core.* Scientists think that Earth's core is made mostly of iron and nickel. The core makes up roughly one-third of Earth's mass. The outer part of the core is molten metal. Because the pressure on the inner core is so high, it consists of very hot solid metal.

Cross Section of Earth

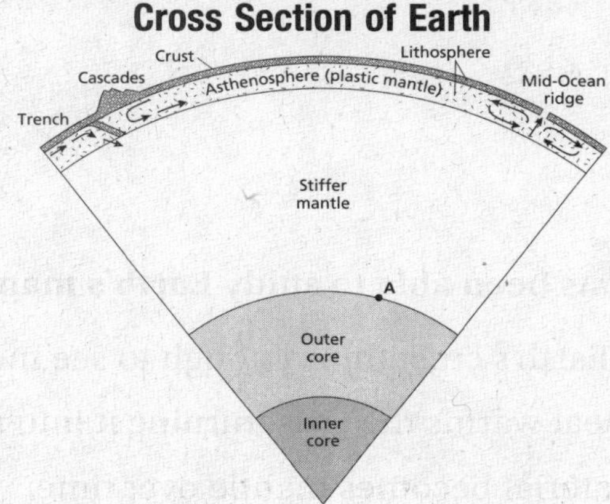

GUIDED PRACTICE

Directions: Using the Standard Review and what you have studied, read each question and circle the letter of the best response.

From what layer of Earth does magma flow from active volcanoes on the ocean's floor?

A crust

B mantle

C inner core

D outer core

The correct answer is B. Magma is molten rock that is pushed through cracks in the crust to form volcanoes. The crust itself (Answer A) is composed of cooler, solid rock. The inner and outer cores (Answers C and D) are composed mostly of iron, not rock, and are far beneath the surface.

Copyright © Holt McDougal. All rights reserved.

STANDARD PRACTICE

1. Where are the least dense compounds found in Earth's layers?

A crust

B mantle

C outer core

D inner core

2. Which layer contains most of Earth's mass?

F continental crust

G oceanic crust

H mantle

J inner core

3. How have scientists been able to study Earth's mantle?

A In some places, Earth's crust thins enough to see mantle.

B Earth's surface heat warms the crust turning it into mantle.

C Cooling crust material becomes mantle over time.

D In some places, mantle rock pushes to Earth's surface.

4. a. How is oceanic crust different from continental crust?

b. Describe what scientists know about Earth's mantle.

Copyright © Holt McDougal. All rights reserved.

GLE 0707.7.4 Explain how earthquakes, mountain building, volcanoes, and sea floor spreading are associated with movements of the earth's major plates.

STANDARD REVIEW

The lithosphere is made up of two parts—the crust and the upper part of the mantle. The lithospheric crust is divided into *tectonic plates,* pieces that move around on the surface. The movement of the tectonic plates is partly a result of the transfer of internal heat from the mantle during sea-floor spreading. *Sea-floor spreading* is the process by which new oceanic lithosphere forms as magma rises toward the surface and solidifies. As the tectonic plates move away from each other, the sea floor spreads apart, and magma fills in the gap. As this new crust forms, the older crust gets pushed away.

As the tectonic plates move across the surface, they collide, releasing large amounts of energy. The kinetic energy of these collisions pushes rock into mountain ranges and causes earthquakes. Volcanoes form at boundaries between plates where the material of the mantle can reach the surface.

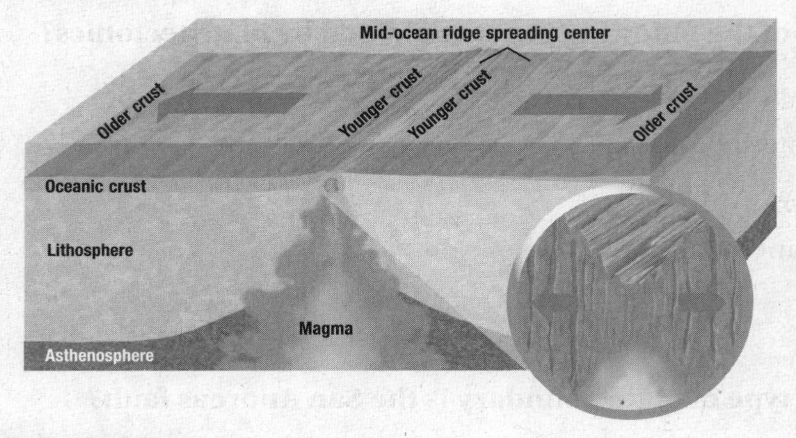

GUIDED PRACTICE

Directions: Using the Standard Review and what you have studied, read each question and circle the letter of the best response.

How do the tectonic plates that border a mid-ocean ridge move in relation to one another?

- **A** Plates move in the same direction.
- **B** The plates move toward one another.
- **C** The two plates move away from one another.
- **D** The plates slide past one another.

The correct answer is C. In mid-ocean ridges, plates move away from one another, allowing the formation of new crust. If they moved in the same direction (Answer A) or slid past one another (Answer D), there would be no crust formation, and, if they moved toward one another (Answer B), material would be pushed up into mountains.

Copyright © Holt McDougal. All rights reserved.

Grade 7 – Earth and Space Science: Standard 7 – The Earth

STANDARD PRACTICE

1. The majority of volcanoes are located in the Pacific Ocean in an area called the Ring of Fire. How is the location of volcanoes in the Ring of Fire related to tectonic plate boundaries?

A Most volcanoes in the Ring of Fire are located at hot spots.

B Most volcanoes in the Ring of Fire are located along a mid-ocean ridge.

C There is no consistent pattern in the location of volcanoes in the Ring of Fire.

D Most volcanoes in the Ring of Fire are located where an oceanic plate collides with a continental plate.

2. Which of the following is <u>not</u> explained by plate tectonics?

F earthquakes

G mountain building

H the layers of Earth

J the supercontinent cycle

3. Which type of plate boundary is the San Andreas fault?

A divergent

B convergent

C transform

D strike-slip

4. a. Would you expect to see a folded mountain range at a mid-ocean ridge? Explain your answer.

b. What happens to crust during subduction?

Copyright © Holt McDougal. All rights reserved.

Grade 7 – Earth and Space Science: Standard 7 – The Earth

GLE 0707.7.5 Differentiate between renewable and nonrenewable resources in terms of their use by man.

STANDARD REVIEW

A *natural resource* is any natural material that is used by humans. Examples of natural resources are water, petroleum, minerals, and forests. Most resources are changed and made into products that make people's lives more comfortable and convenient. The energy we get from resources, such as gasoline and wind, ultimately comes from the sun's energy. Some natural resources can be renewed. A *renewable resource* can be replaced at the same rate at which it is used. Some resources are renewable if they are not used too quickly. Trees, for example, are renewable. However, some forests are being cut down faster than new forests can grow to replace them.

Not all of Earth's natural resources are renewable. A *nonrenewable resource* is a resource that forms at a rate that is much slower than the rate at which it is consumed. When these resources become scarce, humans will have to find other resources to replace them. Most of the energy we use comes from a group of natural resources called fossil fuels. A *fossil fuel* is a nonrenewable energy resource formed from the remains of plants and animals that lived long ago. Examples of fossil fuels include petroleum, coal, and natural gas. Once fossil fuels are used up, new supplies will not be available for thousands—or even millions—of years. To continue to have access to energy and to overcome pollution, we must find alternative sources of energy.

U.S. Energy Sources

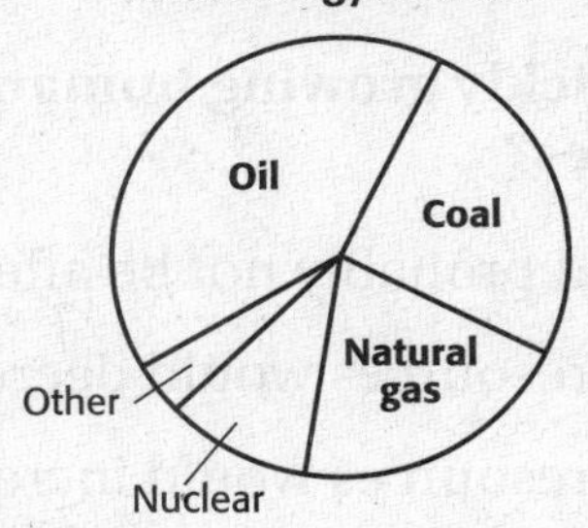

GUIDED PRACTICE

Directions: Using the Standard Review and what you have studied, read each question and circle the letter of the best response.

Which of the following resources is a renewable resource?

A coal **B** trees

C petroleum **D** iron ore

The correct answer is B. Trees are a renewable resource because they can be replaced as quickly as they are used, if forests are properly managed. Fossil fuels, such as coal and petroleum (Answers A and C), renew too slowly to be replaced. Iron ore is a non-renewable resource because it is not replaced in the crust.

Copyright © Holt McDougal. All rights reserved.

Grade 7 – Earth and Space Science: Standard 7 – The Earth

STANDARD PRACTICE

1. Which of the following statements about oil and natural gas reserves is not true?

A Scientists find reserves by using seismic equipment.

B Reserves are generally found under layers of impermeable rock.

C Oil and natural gas are renewable resources.

D Extracting reserves usually requires drilling through rock.

2. Which of the following is true of renewable resources?

F They must be converted into nonrenewable resources.

G They are less useful than nonrenewable resources.

H Many of them can become scarce if used too quickly.

J No matter how much we conserve, they will one day be gone.

3. What effects might a quickly growing human population have on the world's natural resources?

A Natural resources would probably not be affected.

B The amount of natural resources would decrease.

C The amount of natural resources would increase.

D The amount of natural resources that would be recycled would decrease.

4. a. Why do we need to conserve renewable resources even though they can be replaced?

b. How can the use of the renewable resource, hydroelectric power, have undesirable effects?

Copyright © Holt McDougal. All rights reserved.

Grade 7 – Earth and Space Science: Standard 7 – The Earth

GLE 0707.7.6 Evaluate how human activities affect the earth's land, oceans, and atmosphere.

STANDARD REVIEW

Human activities can have harmful effects on the environment. Farming and cutting timber can damage soil if these activities are not done carefully. When soil is left unprotected, it can be exposed to erosion. *Erosion* is the process by which wind, water, ice, or gravity transport soil and sediment from one location to another. Roots from plants and trees are like anchors to the soil. Roots keep topsoil from being eroded. Therefore, plants and trees protect the soil. By taking care of the vegetation, you also take care of the soil.

Mining gives us the minerals we need, but it may also create problems. Mining can destroy or disturb the habitats of plants and animals. Waste products from a mine may get into water sources and pollute surface water and groundwater.

Atmospheric data show that average global temperatures have increased in the past 100 years. Such an increase in average global temperatures is called *global warming.* Some scientists have hypothesized that an increase of greenhouse gases in the atmosphere has caused this warming trend. *Greenhouse gases* are gases that absorb thermal energy in the atmosphere. Human activity, such as the burning of fossil fuels and deforestation, have increased levels of greenhouse gases, such as carbon dioxide, in the atmosphere. If this hypothesis is correct, increasing levels of greenhouse gases will cause average global temperatures to continue to rise. Disrupted global climate patterns would affect plants and animals adapted to live in specific climates.

GUIDED PRACTICE

Directions: Using the Standard Review and what you have studied, read each question and circle the letter of the best response.

Poor soil conservation practices will

A not affect farmers, as topsoil is an inexhaustible resource.

B exhaust Earth's limited supply of topsoil.

C result in the expensive process of manufacturing new replacement topsoil.

D result in excessive oxidation of the soil's humus content.

The correct answer is B. If soil is not conserved, the nutrient-rich topsoil becomes depleted and can even be transported away by wind and water. Soil renews very slowly, so it is not an inexhaustible resource (Answer A), and it cannot be manufactured (Answer C). Oxidation of humus does not harm soil (Answer D).

Copyright © Holt McDougal. All rights reserved.

Grade 7 – Earth and Space Science: Standard 7 – The Earth

STANDARD PRACTICE

1. A scientist compiled data from numerous field investigations to create the pie chart showing the various causes of soil damage. What percentage of soil damage is caused by wind and water erosion?

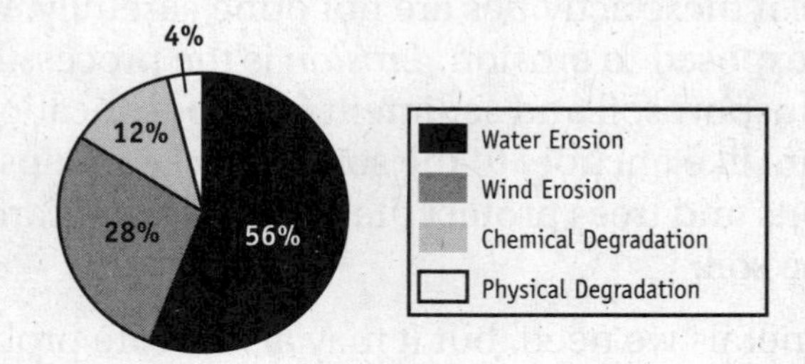

A 16%

B 28%

C 56%

D 84%

2. What is one result of humans burning large amounts of fossil fuels?

F soil erosion

G abrasion

H contour plowing

J acid precipitation

3. Which of these is an indication of soil overuse?

A low nutrient levels

B high pH

C too much organic material

D sandy texture

4. a. Explain why removing bordering forests results in siltation of adjacent streams.

b. How does siltation of streams affect aquatic wildlife?

Copyright © Holt McDougal. All rights reserved.

Grade 7 – Physical Science: Standard 11 – Motion

GLE 0707.11.1 Identify six types of simple machines.

STANDARD REVIEW

A *machine* is a device that makes work easier by changing the amount of force applied to cause a motion, or the distance of the motion, or both. Machines can also change the direction in which a motion occurs. The six simple machines are the lever, the inclined plane, the wedge, the screw, the pulley, and the wheel and axle. All machines are made from one or more of these simple machines.

A *lever* is a simple machine that has a bar that pivots at a fixed point, called a *fulcrum*. Levers are used to apply a force to a load. The lever changes the direction of motion, the distance of the motion, or both.

A *pulley* is a simple machine that has a grooved wheel that holds a rope or a cable. A load is attached to one end of the rope, and an input force is applied to the other end.

A *wheel and axle* is a simple machine consisting of two circular objects of different sizes that turn together. Turning the wheel results in a mechanical advantage of greater than 1 because the radius of the wheel is larger than the radius of the axle.

An *inclined plane* is a simple machine that is a straight, slanted surface. A ramp is an inclined plane. The inclined plane reduces the force needed to raise an object by increasing the distance.

A *wedge* is a pair of inclined planes that move. A wedge applies an output force that is greater than your input force, but you apply the input force over a greater distance.

A *screw* is an inclined plane that is wrapped in a spiral around a cylinder. When a screw is turned, a small force is applied over the long distance along the inclined plane of the screw. Meanwhile, the screw applies a large force through the short distance it is pushed.

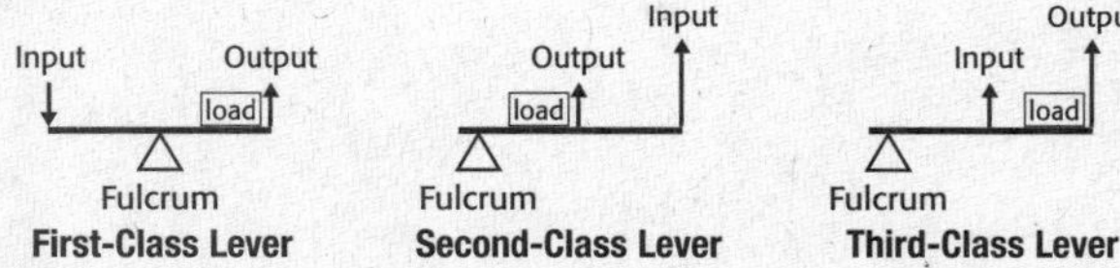

GUIDED PRACTICE

Directions: Using the Standard Review and what you have studied, read each question and circle the letter of the best response.

How does a second-class lever affect the force and direction of motion?

A changes force, but not direction

B changes force and direction

C changes direction, but not force

D does not change either force or direction

The correct answer is A. In a second class lever the load moves in the same direction as the input force, but the amount of force required to move the load is decreased. Answers B and C are incorrect because direction does not change. Answers C and D are incorrect because the force does change.

Copyright © Holt McDougal. All rights reserved.

Grade 7 – Physical Science: Standard 11 – Motion

STANDARD PRACTICE

1. How does the third-class lever, shown on the previous page, make work easier?

A changing the direction of the force

B increasing both force and distance

C increasing force and decreasing distance

D decreasing force and increasing distance

2. Which of the following machines acts as a third-class lever?

F seesaw

G bottle opener

H wheelbarrow

J arm lifting a barbell

3. Which of these tools is an example of a wedge?

A loading ramp

B knife

C hammer

D doorknob

4. a. How is a screw related to an inclined plane?

b. How does a screw increase the force that is applied to it?

Copyright © Holt McDougal. All rights reserved.

Grade 7 – Physical Science: Standard 11 – Motion

GLE 0707.11.2 Apply the equation for work in experiments with simple machines to determine the amount of force needed to do work.

STANDARD REVIEW

$$W = f \times d$$

The equation above shows the relationship of work (*W*), force (*f*), and distance (*d*). It also explains why machines are useful. *Work* is calculated by multiplying the amount of force applied by the distance of motion in the direction of the force. If you want to lift a heavy box onto a 1 meter high platform, you may need to apply more force than your body is able to exert. However, the work to lift the box is the same whether you lift it 1 meter or push it up a 4 meter ramp. The ramp increases the distance by a factor of 4, but it also decreases the force needed by the same factor. The input work and the output work are the same, so force decreases when distance increases.

The *mechanical advantage* of a machine is defined as the output force divided by the input force or the input distance divided by the output distance. The 4 meter ramp has a mechanical advantage of 4. If the machine increases the distance, less input force is needed, and the mechanical advantage is greater than 1. Less force is needed, but you only get 1 meter of elevation for 4 meters of pushing. Sometimes machines work by increasing the output distance. The wheel and axle of a car reduce the force applied by the engine, but the wheel moves a greater distance than the axle each time it turns. The mechanical advantage is less than 1. If there is a change in direction, but not in distance or force, the mechanical advantage is equal to 1.

$$\textbf{Mechanical Advantage} = \frac{\textbf{output force}}{\textbf{input force}} = \frac{\textbf{input distance}}{\textbf{output distance}}$$

GUIDED PRACTICE

Directions: Using the Standard Review and what you have studied, read each question and circle the letter of the best response.

A first-class lever is used to lift a 1,000 N rock. The free end of the lever is pushed downward a distance of 1 meter, but the rock is raised only 10 centimeters. How much force was exerted on the lever to raise the rock?

A 10 N **B** 100 N

C 1,000 N **D** 10,000 N

The correct answer is B. The mechanical advantage of the lever is 10 because the end of the lever moved 10 times as far as the rock. The output distance decreased by a factor of 10 so the output force increased by a factor of 10, and the input force was 100 N. Answer A would correspond to a mechanical advantage of 100. For Answer C to be correct, the mechanical advantage would have to be 1, and the output distance would be equal to the input distance. Answer D indicates that the mechanical advantage is 0.1, so the rock would move farther than the free end of the lever.

Copyright © Holt McDougal. All rights reserved.

Grade 7 – Physical Science: Standard 11 – Motion

STANDARD PRACTICE

1. Which of these machines can have a mechanical advantage equal to 1?

A screw

B third-class lever

C pulley

D wheel and axle

2. What is the mechanical advantage of the block and tackle system shown in the illustration to the right?

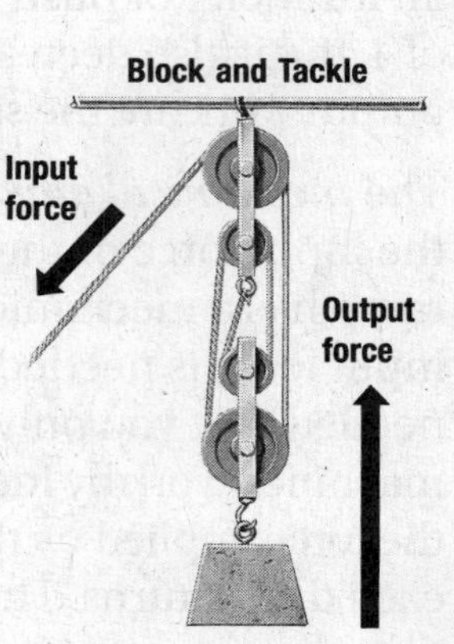

F 1

G 4

H 5

J cannot be determined from this information

3. For which of these simple machines is the output force always less than the input force?

A first-class lever

B second-class lever

C third-class lever.

D wedge

4. a. How does the mechanical advantage of a machine affect its work output?

b. How does the work output of a real machine compare to the work input of the machine? Explain your answer.

Copyright © Holt McDougal. All rights reserved.

GLE 0707.11.3 Distinguish between speed and velocity.

STANDARD REVIEW

The movement of an object is always measured by comparing its position to that of a stationary *reference point.* When an object changes position over time relative to a reference point, the object is in *motion.* Earth's surface is a common reference point for determining motion.

Speed is the distance traveled by an object divided by the time taken to travel that distance. The speedometer of a car shows how fast the car is moving at any particular time. When you calculate the speed of an object, its direction does not matter, only the distance and the time. *Velocity* is the speed of an object in a particular direction. Be careful not to confuse the terms *speed* and *velocity.* They do not have the same meaning. Velocity must include a reference direction. If you say that an airplane's velocity is 600 km/h, you would not be correct. However, you could say that the plane's velocity is 600 km/h south. You can think of velocity as the rate of change of an object's position. An object's velocity is constant, only if its speed and direction do not change. Therefore, constant velocity is always motion along a straight line. An object's velocity changes if either its speed or direction changes.

Acceleration is the rate at which velocity changes. Velocity changes if speed changes, if direction changes, or if both change. So, an object accelerates if its speed, its direction, or both change. Acceleration depends not only on how much velocity changes, but also on how fast velocity changes. The faster the velocity changes, the greater the acceleration.

GUIDED PRACTICE

Directions: Using the Standard Review and what you have studied, read each question and circle the letter of the best response.

A satellite orbiting Earth at a constant distance moves at a constant speed. Which statement about its velocity is accurate?

A The velocity of the satellite is constant.

B The velocity is changing at all times.

C There is too little information to know whether the velocity is constant.

D Velocity is constant based on some reference points and changing based on other reference points.

The correct answer is B. Velocity depends on speed and direction. Because the direction is constantly changing, velocity is constantly changing. An object with constant velocity (Answer A) moves in a straight line. The direction is constantly changing based on any reference point (Answer D). That is the only information that you need to know that velocity is constantly changing (Answer C).

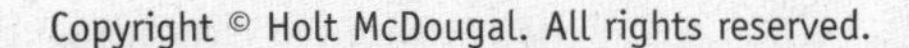

Copyright © Holt McDougal. All rights reserved.

STANDARD PRACTICE

1. Plane A is traveling 400 km/h west compared to a reference point on the ground. Plane B is also moving 400 km/h west compared to the same reference point, but it is 100 miles east of Plane A. Using Plane A as a reference point, what is the velocity of Plane B?

A 0 km/h

B 100 km/h west

C 100 km/h east

D 400 km/h west

2. What causes an object to accelerate?

F a change in its speed

G a change in direction of motion

H a change in velocity

J all of the above

3. A car is traveling 30 km/h west. A second car is traveling 20 km/h west. What is the speed of the second car relative to the first.

A 10 km/h

B 50 km/h

C 10 km/h west

D 10 km/h east

4. a. How can the speed of an object vary depending on the reference point?

b. Is it possible for two objects to have the same velocity, but different speeds? Explain your answer.

Copyright © Holt McDougal. All rights reserved.

Grade 7 – Physical Science: Standard 11 – Motion

GLE 0707.11.4 Investigate how Newton's Laws of Motion explains an object's movement.

STANDARD REVIEW

Newton's First Law of Motion
An object at rest remains at rest, and an object in motion remains in motion at constant speed and in a straight line unless acted on by an unbalanced force.
Newton's Second Law of Motion
The acceleration of an object depends on the mass of the object and the amount of force applied.
Newton's Third Law of Motion
Whenever one object exerts a force on a second object, the second object exerts an equal and opposite force on the first.

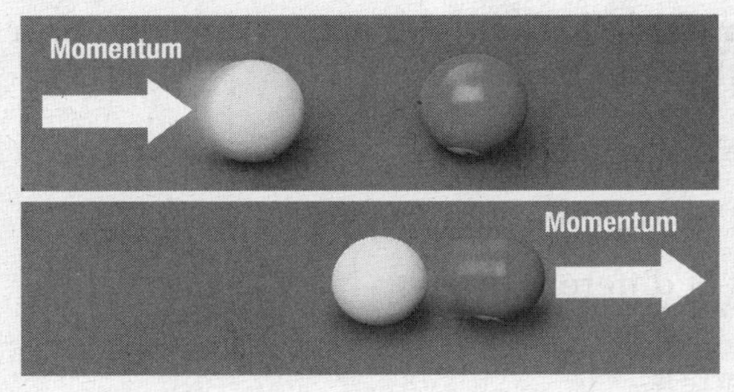

The three laws of motion describe what happens when the balls collide. The light ball will continue moving and the dark ball will remain stationary, unless a force acts on them (First Law). During a collision, the light ball exerts a force that accelerates the dark ball and causes the dark ball to move (Second Law). The dark ball exerts an equal, and opposite, force on the light ball, causing the light ball to stop moving (Third Law).

GUIDED PRACTICE

Directions: Using the Standard Review and what you have studied, read each question and circle the letter of the best response.

In the collision shown above, what forces are exerted during the collision.

A Only the light ball exerts a force because it is moving.

B Only the dark ball exerts a force because its motion changes.

C Each ball exerts a force on the other, but the moving ball exerts a greater force.

D Each ball exerts an equal force on the other ball.

The correct answer is D. According to Newton's third law, each ball exerts a force that is equal in magnitude and opposite in direction (Answer C). The light ball exerts a force that changes the velocity of the dark ball while at the same time the dark ball exerts a force that changes the velocity of the light ball (Answers A and B).

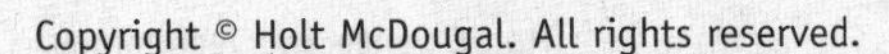
Copyright © Holt McDougal. All rights reserved.

Grade 7 – Physical Science: Standard 11 – Motion

STANDARD PRACTICE

1. Two forces are acting on an object, but the net force on the object is 0 N. For the net force to be 0 N, all the forces on the object must cancel. What must be true for the two forces on the object to cancel?

A The forces are the same size and in the same direction.

B The forces are different sizes and in the same direction.

C The forces are the same size and in opposite directions.

D The forces are different sizes and in opposite directions.

2. When a soccer ball is kicked, the action and reaction forces do not cancel each other out because

F the forces are not equal in size.

G the forces act on different objects.

H the forces act at different times.

J all of the above

3. A golf ball and a bowling ball are moving at the same velocity. Which of the two has more momentum?

A The golf ball has more momentum because it has less mass.

B The bowling ball has more momentum because it has more mass.

C They have the same momentum because they have the same velocity.

D There is not enough information to determine the answer.

4. During a space shuttle launch, about 830,000 kg of fuel is burned in 8 min. The fuel provides the shuttle with a constant thrust, or forward force.

a. How does Newton's Third Law of Motion explain why the shuttle's acceleration increases as the fuel is burned?

b. What happens to the shuttles acceleration when the engines turn off?

Copyright © Holt McDougal. All rights reserved.

Grade 7 – Physical Science: Standard 11 – Motion

GLE 0707.11.5 Compare and contrast the basic parts of a wave.

STANDARD REVIEW

The *amplitude* of a wave is related to its height. For a mechanical wave, the amplitude is the maximum distance that the particles of a medium vibrate from their rest position. The larger the amplitude is, the taller the wave. A wave with a larger amplitude carries more energy than a wave with a smaller amplitude.

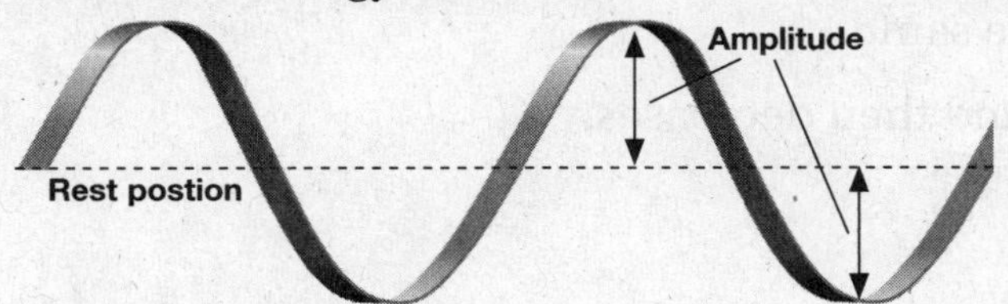

The *wavelength* of a wave is the distance between any two crests or compressions next to each other in a wave. The distance between two troughs or rarefactions next to each other is also equal to one wavelength. The wavelength can be measured from any point on a wave to the corresponding point on the next wave. A wave with a shorter wavelength carries more energy than a wave with a longer wavelength does if they have the same amplitude. The figure below shows how to measure wavelength on a transverse wave.

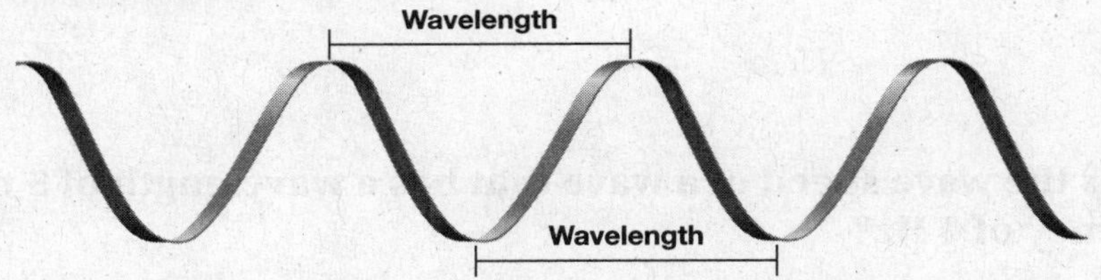

The number of waves produced in a given amount of time is the *frequency* of the wave. Frequency is usually expressed in *hertz* (Hz). One hertz equals one wave per second. If the amplitudes are equal, higher frequency waves carry more energy than low frequency waves.

Wave speed is the speed at which a wave travels. Wave speed can by calculated by multiplying the wavelength by the frequency.

GUIDED PRACTICE

Directions: Using the Standard Review and what you have studied, read each question and circle the letter of the best response.

Which of the following results in a higher energy wave?

A shorter wavelength **B** lower frequency

C smaller amplitude **D** lower speedThe correct answer is A.

For waves with the same amplitude, more energy is carried by a wave with a shorter wavelength. A lower frequency (Answer B) corresponds to a wave with a longer wavelength and lower energy. Energy decreases with a decrease in amplitude (Answer C). A slower wave (Answer D) has a lower frequency and less energy.

Copyright © Holt McDougal. All rights reserved.

STANDARD PRACTICE

1. As the wavelength increases, the frequency

A decreases.

B increases.

C remains the same.

D increases and then decreases.

2. The wave property that is related to the height of a wave is the

F wavelength.

G amplitude.

H frequency.

J wave speed.

3. What is the wave speed of a wave that has a wavelength of 5 m and a frequency of 4 Hz?

A 0.8 m/s

B 1.25 m/Hz

C 9 m/Hz

D 20 m/s

4. Imagine that you are standing by the edge of a pond. A leaf is floating in the middle of the pond. You throw a rock into the pond, and it hits the water near the leaf. The waves created by the rock entering the water cause the leaf to bob up and down.

a. What causes the leaf to bob up and down?

b. Could you make the leaf float to the edge of the pond by throwing several rocks on one side of the leaf? Explain your answer.

Copyright © Holt McDougal. All rights reserved.

Grade 7 – Physical Science: Standard 11 – Motion

GLE 0707.11.6 Investigate the types and fundamental properties of waves.

STANDARD REVIEW

Most waves transfer energy by the vibration of particles in a medium. A *medium* is a substance through which a wave can travel. A medium can be a solid, a liquid, or a gas. When a particle vibrates, it can pass its energy to a particle next to it. The second particle will then vibrate and pass energy to another particle and so on throughout the medium.

Sound waves are longitudinal waves that need a medium. Sound energy travels by the vibration of particles. If there are no particles to vibrate, no sound is possible. If you put an alarm clock inside a jar and remove all the air from the jar to create a vacuum, you will not be able to hear the alarm. Other waves that need a medium include ocean waves, which move through water, vibrations of guitar and cello strings when they vibrate, and seismic waves that travel through Earth. Waves that need a medium are called *mechanical waves.*

Electromagnetic waves are generated by the interaction of electric and magnetic fields. They transfer energy without a medium so they are able to pass through a vacuum. Visible light is the most familiar example of electromagnetic waves. Others include microwaves, TV and radio signals, and the X rays used by dentists and doctors.

All waves transfer energy as they travel, and waves can do work. The energy of mechanical waves is carried by vibrations of particles. The energy of an electromagnetic wave is carried by the interaction of magnetic and electrical forces.

GUIDED PRACTICE

Directions: Using the Standard Review and what you have studied, read each question and circle the letter of the best response.

How does a wave transfer energy through a medium?

A The wave gains energy from the particles in the medium.

B The wave moves between the particles of the medium.

C The particles in the medium vibrate and collide with one another.

D The particles in the medium move in the direction of the wave.

The correct answer is C as vibrating particles collide, they transfer energy from one to another. Answer A is incorrect because the particles gain energy from the wave. The wave does not move between particles (Answer B); it is the motion of the particles. Particles vibrate but do not move with the wave (Answer D).

Copyright © Holt McDougal. All rights reserved.

STANDARD PRACTICE

1. What kind of waves travel through Earth?

A seismic waves

B radio waves

C light waves

D water waves

2. Electromagnetic waves are different from other types of waves because they can travel through

F air.

G glass.

H space.

J steel.

3. Which of the following electromagnetic waves have the longest wavelengths?

A X rays

B visible light waves

C gamma rays

D infrared waves

4. The sun produces a large amount of energy. Some of that energy is carried away from the sun in the form of electromagnetic waves. A small amount of the electromagnetic waves from the sun travel through space to Earth every day. The electromagnetic waves from the sun include infrared waves, visible light, and ultraviolet light.

a. Explain the difference between infrared waves, visible light, and ultraviolet light.

b. Why can electromagnetic waves from the sun travel to Earth?

Copyright © Holt McDougal. All rights reserved.

TCAP Test Preparation
Practice Test A

1. The pictures below show the textures of four igneous rocks that were collected during a field investigation. These rocks formed from magma that cooled at various rates. The faster magma cools, the smaller the crystals form in the rock. Which of the rocks formed from magma that cooled very rapidly?

A syenite

B granite

C obsidian

D pegmatite

2. The model that is <u>most</u> useful in studies of global warming is a

F physical model of the ocean.

G mathematical model of the ocean.

H physical model of the atmosphere.

J mathematical model of the atmosphere.

Copyright © Holt McDougal. All rights reserved.

3. One of Newton's laws states that an object traveling at a constant speed in a specific direction will continue to do so, unless an unbalanced force acts on it. The moon orbits Earth because

A no unbalanced force acts on it.

B an unbalanced gravitational force constantly pulls the moon toward Earth.

C circular forces act on it.

D inertia pulls the moon toward Earth.

4. Which of the following statements regarding light is true?

F Light is a form of electrical energy that travels through wires.

G Light is a form of energy that travels as a wave and can pass through different media.

H Light is generated by the sun and is captured for human use.

J Light is a process by which energy is generated.

Copyright © Holt McDougal. All rights reserved.

5. Maria is about to conduct a field investigation on plate tectonics. She wants to study a region with transform boundaries. Look at the diagram. Which plate boundary would be a good choice for Maria to study?

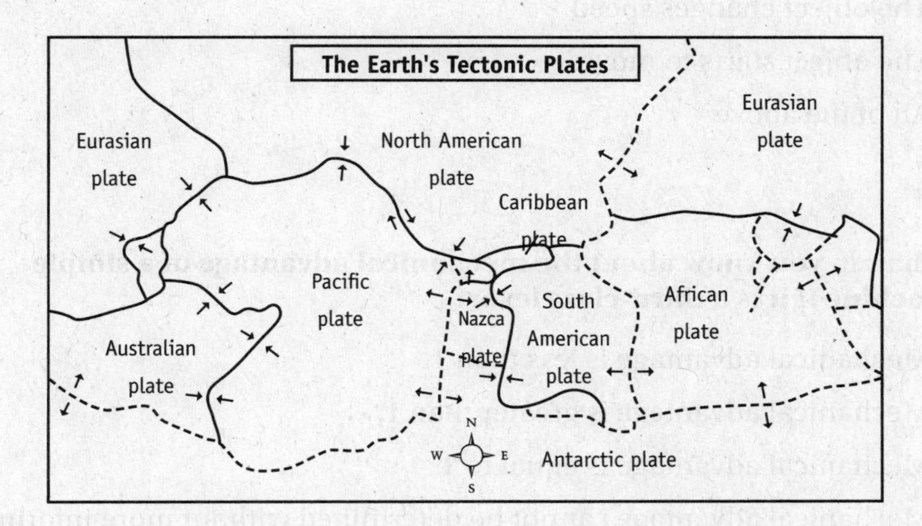

A Eurasian plate and North American plate

B Pacific plate and North American plate

C African plate and Antarctic plate

D South American plate and Nazca plate

6. What is the purpose of meiosis in cells?

F prepare cells for sexual reproduction

G prepare cells for asexual reproduction

H enable an organism to grow

J repair damaged cells

Copyright © Holt McDougal. All rights reserved.

7. Which of the following may happen when an object receives unbalanced forces?

A The object changes direction.

B The object changes speed.

C The object starts to move.

D All of the above

8. What do you know about the mechanical advantage of a simple machine if it is a third-class lever?

F Mechanical advantage is less than 1.

G Mechanical advantage is greater than 1.

H Mechanical advantage is equal to 1.

J Mechanical advantage cannot be determined without more information.

Copyright © Holt McDougal. All rights reserved.

9. Rushmi constructed the model of Earth shown. What would be the correct order of layers from the outside of Earth to the inside?

Interior of the Earth

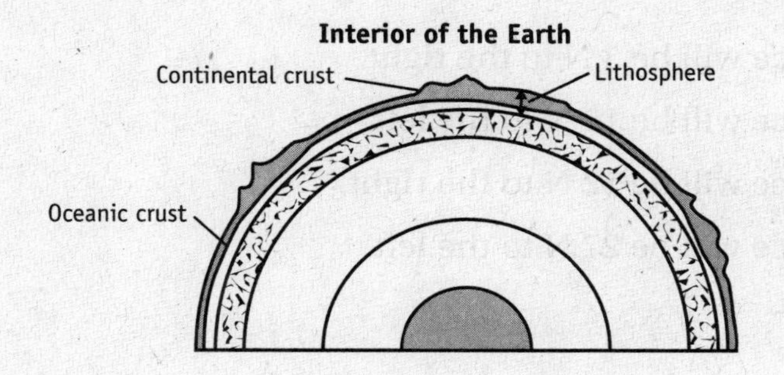

A outer core, inner core, crust, mantle

B crust, mantle, inner core, outer core

C crust, mantle, outer core, inner core

D mantle, outer core, inner core, crust

10. Which simple machine can be represented as an inclined plane wrapped around a cylinder?

F wedge

G screw

H wheel and axle

J pulley

Copyright © Holt McDougal. All rights reserved.

11. A teenager pulls a rope to the left with a force of 12 N. A child pulls on the other end of the rope to the right with a force of 7 N. The child's friend adds a force of 8 N, also pulling to the right. What will happen?

A The net force will be 3 N to the right.

B The net force will be 15 N to the left.

C The net force will be 12 N to the right.

D The net force will be 27 N to the left.

12. In a vacuum, all types of electromagnetic waves have the same

F speed.

G pitch.

H wavelength.

J amplitude.

Copyright © Holt McDougal. All rights reserved.

13. The table below shows the possible combinations of alleles for fur color in rabbits. What is the dominant trait for fur color in rabbits?

Genotype	Phenotype
BB	Black fur
Bb	?
bb	White fur

A white fur
B brown fur
C black fur
D gray fur

14. A place within Earth where the speed of seismic waves increases sharply is the

F shadow zone.
G inner core.
H moho.
J epicenter.

Copyright © Holt McDougal. All rights reserved.

15. Which of the following is not a renewable resource?

A coal

B solar energy

C wind energy

D geothermal energy

16. A wave has a wavelength of 3 m and a frequency of 5 Hz. What is the wave's speed?

F 0.6 m/s

G 5 m/s

H 12 m/s

J 15 m/s

Copyright © Holt McDougal. All rights reserved.

17. As part of a laboratory experiment, a student crosses a true-breeding tall plant having alleles of (TT) with another plant having alleles of (Tt). Which of the following Punnett squares correctly shows the cross?

A

	T	T
t	Tt	Tt
t	Tt	Tt

B

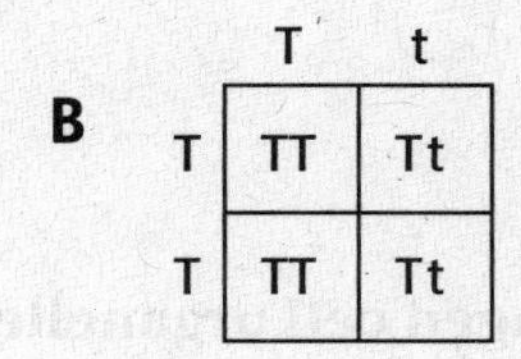

C

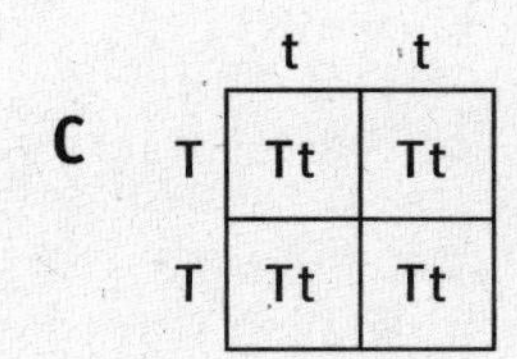

D

	T	t
T	TT	Tt
t	Tt	tt

18. Which of the following is <u>not</u> a possible cause of tectonic plate motion?

F convection

G compression

H ridge push

J slab pull

Copyright © Holt McDougal. All rights reserved.

19. How did Wegener's hypothesis about continental drift help to advance science, even though it was not immediately accepted?

A It advanced science only because it was true.

B It supported the accepted hypotheses about Earth's structure.

C It suggested ideas for new types of experiments that increased understanding of Earth.

D It advanced science by making Wegener famous so he could do more research.

20. A cell vesicle containing enzymes that destroy damaged cell organelles is called

F an endoplasmic reticulum

G a mitochondrion

H a Golgi complex

J a lysosome

Copyright © Holt McDougal. All rights reserved.

21. What is the relationship between major mountain ranges in California and the plate boundary?

- **A** The mountain ranges are oriented perpendicular to the plate boundary.
- **B** The mountain ranges are oriented parallel to the plate boundary.
- **C** The mountain ranges are oriented on the plate boundary.
- **D** The orientation of the mountain ranges is changed by the plate boundary.

22. When people practice conservation, how do they affect the environment?

- **F** They do not affect the environment.
- **G** They protect the environment by using fewer natural resources.
- **H** They decrease the amount of natural resources in the environment.
- **J** They cause an increase in the amount of pollution in the environment.

Copyright © Holt McDougal. All rights reserved.

23. Which of the following strategies would <u>not</u> help conserve the environment?

A reduce pesticide use

B develop alternative energy sources

C protect habitats

D increase the number of landfills

24. Two labs examine a tool found in a cave. One lab determines that it was made about 4,000 years ago, and the second lab determines that it was made 4,800 years ago. What should the researchers conclude about these data?

F The difference is only 800 years, so it does not matter.

G One of the labs made a mistake in their analysis.

H More investigation is needed to determine why the values are different.

J The tool was made 4,400 years ago.

Copyright © Holt McDougal. All rights reserved.

25. **The graph below shows the composition (by weight) of municipal solid wastes in the United States.**

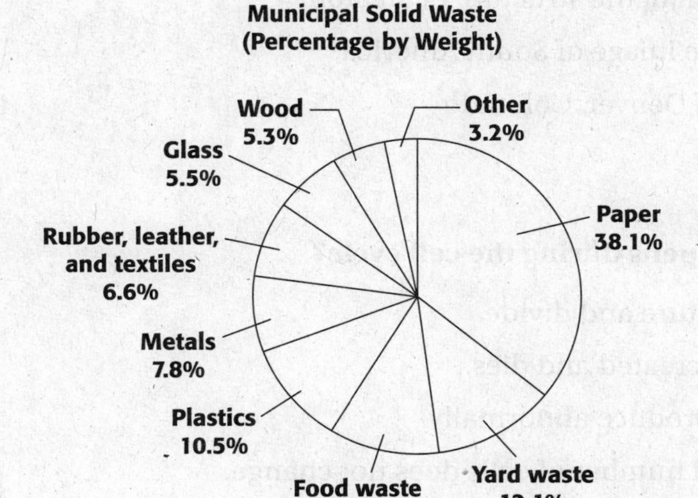

By how much could you conclude that the amount of municipal waste would be reduced if every household in the country had a compost pile?

A 12%

B 23%

C 35%

D 41%

26. **How is photosynthesis important to consumers?**

F Consumers use oxygen and glucose produced by photosynthesis.

G Consumers supply the light energy needed for photosynthesis.

H Consumers supply the water needed for photosynthesis.

J Consumers supply the chloroplasts to producers.

Copyright © Holt McDougal. All rights reserved.

27. Which of the following is an example of a scientist's physical model?

A a crash-test dummy for a car company

B a diagram of the structure of an atom

C a satellite image of South America

D a map of Denver, Colorado

28. What happens during the cell cycle?

F Cells mature and divide.

G A cell is created and dies.

H Cells reproduce abnormally.

J The total number of cells does not change.

Copyright © Holt McDougal. All rights reserved.

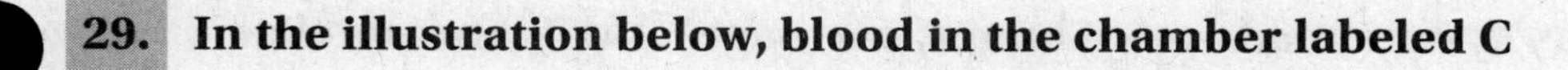

29. In the illustration below, blood in the chamber labeled C

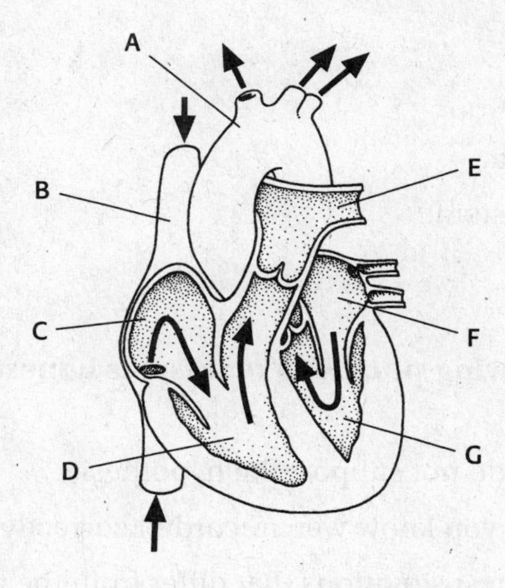

A is full of oxygen.

B is oxygen-poor.

C has very little plasma.

D is returning from the lungs.

30. Which of the following statements about asexual reproduction is <u>not</u> true?

F The offspring are copies of the parent.

G It results in more variation in species than does sexual reproduction.

H Only one parent sex cell is needed.

J Most single-celled organisms reproduce this way.

Copyright © Holt McDougal. All rights reserved.

31. The body system made up of bones, cartilage, and ligaments that supports your body and allows you to move is called the

A skeletal system.

B endocrine system.

C respiratory system.

D cardiovascular system.

32. Which of the following situations represents honest, clear, and accurate record keeping?

F erasing data that do not support the hypothesis

G revising data that you know were recorded correctly

H making a note of test situations that differ from the procedure

J fixing numerical observations so that they match expected results

Copyright © Holt McDougal. All rights reserved.

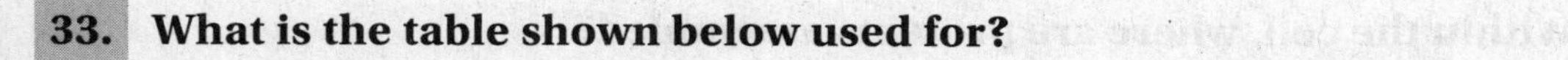

33. **What is the table shown below used for?**

	p	*p*
P	*Pp*	*Pp*
P	*Pp*	*Pp*

A to show recessive traits

B to show dominant traits

C to show homozygous alleles

D to predict genetic cross results

34. **Why is it important for scientists to write careful reports explaining their observations and measurements?**

F so that no one can untruthfully claim to have made a discovery first

G so that others can repeat their experiments and verify their results

H so that the hypotheses can be quickly formulated into theories

J so that they can prove that they used a scientific method

Copyright © Holt McDougal. All rights reserved.

35. Within the cell, where are proteins assembled?

A in the nucleus

B in the cytoplasm

C in the amino acids

D on the chromosomes

36. Which of the following is found inside the nucleus of a cell?

F DNA

G ribosomes

H the cytoskeleton

J the Golgi complex

Copyright © Holt McDougal. All rights reserved.

37. Which explanation, for the chart below, best describes why the author chose to present these data in a pie chart?

Threatened and Endangered Amphibian Species in the United States

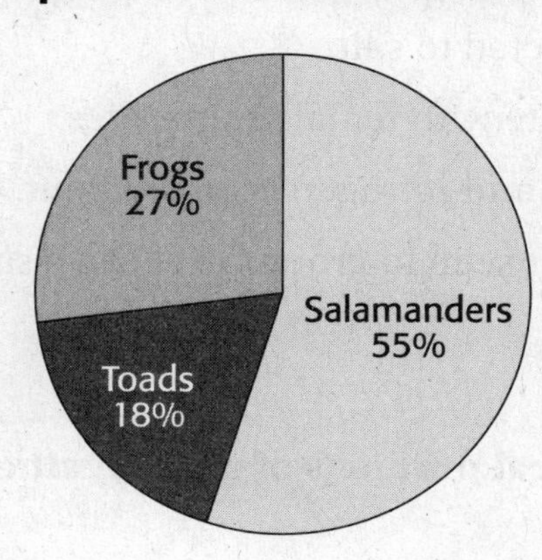

Source: U.S. Fish and Wildlife Service.

A He wanted to show how amphibians have become more endangered from one year to the next.

B He wanted to show how amphibians have become more endangered continuously over time.

C He wanted to compare the percentages of each type of amphibian that is endangered or threatened.

D He wanted to show that all amphibians are threatened or endangered.

38. A circular pie chart has a wedge that is 45% of the chart. If there are 360º in a circle, how many degrees are there in the wedge?

F 45°

G 162°

H 198°

J 315°

Copyright © Holt McDougal. All rights reserved.

39. Jane and Jim observed a group of male butterflies by the roadside. As an experiment, and in an area where butterflies had been seen, they put out different trays full of sand soaked in saltwater and sand soaked in a solution containing nitrogen. Which of the following statements is a testable hypothesis about the experiment?

A Butterflies are attracted to salt.

B Male butterflies mate with female butterflies.

C Salt is a compound and nitrogen is an element.

D Butterflies are never seen in groups except on sandy surfaces.

40. What are the chemical products of photosynthesis?

F glucose and water

G sucrose and carbon dioxide

H water and carbon dioxide

J oxygen and glucose

Copyright © Holt McDougal. All rights reserved.

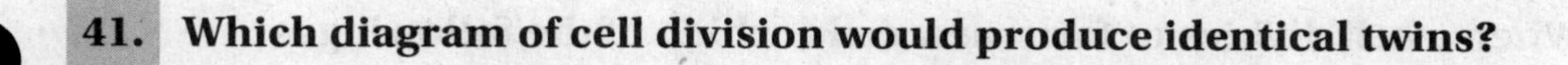

41. Which diagram of cell division would produce identical twins?

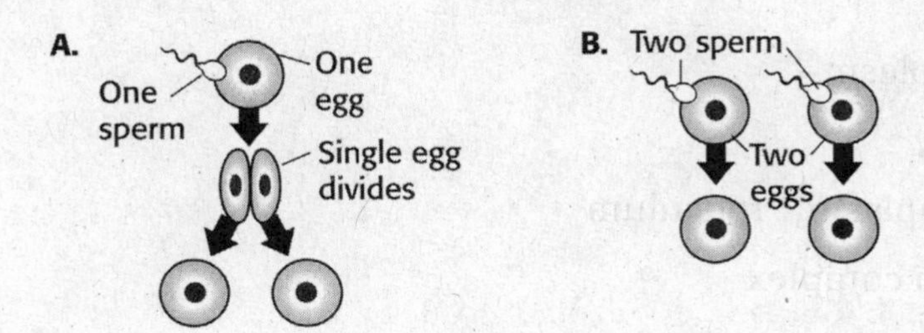

A diagram A, because a single fertilized egg separates into two halves

B diagram B, because each egg is fertilized by a separate sperm cell

C both diagram A and diagram B, because twins result in both cases

D diagram B, because two eggs are released by an ovary

42. Which of the following organisms use the process of cellular respiration?

F animals

G plants

H bacteria

J all of the above

Copyright © Holt McDougal. All rights reserved.

43. When a cell divides to form new cells, which of the following is responsible for carrying the genetic material of parent cells to the new cells?

A the cytoplasm
B the DNA
C the endoplasmic reticulum
D the Golgi complex

44. Which of the following is not a level of organization in a multicellular organism?

F organ
G organ system
H prokaryotic cell
J tissue

Copyright © Holt McDougal. All rights reserved.

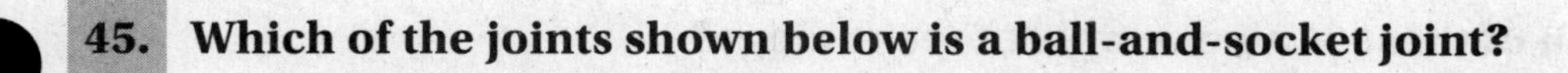

45. Which of the joints shown below is a ball-and-socket joint?

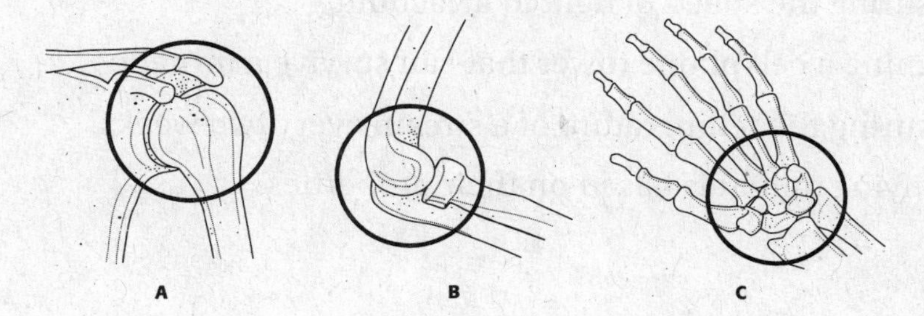

A A

B B

C C

D A, B, and C

46. When pea plants reproduce, two individual parent cells join together to form a new pea plant (offspring). What type of reproduction is involved in producing new pea plants?

F sexual reproduction

G asexual reproduction

H meiosis

J mitosis

Copyright © Holt McDougal. All rights reserved.

47. Which of these is an example of technology?

A measuring the speed of light in a vacuum

B designing a cell phone tower that can survive a tornado

C measuring the temperature of a stream every two weeks

D classifying minerals based on their properties

48. Blood is made up of red blood cells, white blood cells, platelets, and plasma. Red and white blood cells are very different in their appearance. For example, white blood cells have a nucleus, but red blood cells do not. What can you conclude about the two types of blood cells based on their different structures?

F They have different functions.

G They have the same function.

H Their structure is not related to their function.

J Red blood cells are prokaryotic, and white blood cells are not.

Copyright © Holt McDougal. All rights reserved.

49. The illustrations below show two types of reproduction. Which of the following statements correctly summarizes the types of reproduction and the related diversity of the offspring?

A Illustration A: sexual reproduction, diverse offspring
Illustration B: asexual reproduction, diverse offspring

B Illustration A: asexual reproduction, diverse offspring
Illustration B: asexual reproduction, offspring same as parent

C Illustration A: sexual reproduction, diverse offspring
Illustration B: asexual reproduction, offspring same as parent

D Illustration A: sexual reproduction, offspring same as parent
Illustration B: sexual reproduction, diverse offspring

50. A white flower that has the genotype *pp* is crossed with a purple flower that has the genotype *PP*. What are the possible genotypes of the offspring?

F *PP* and *pp*

G *PP*

H *Pp*

J *pp*

Copyright © Holt McDougal. All rights reserved.

51. Which of the following is not a step in the engineering design process?

A building models and prototypes

B ignoring failed experiments

C testing new materials

D evaluating test results

52. Which of the following pairs of terms does not represent a system followed by a part of that system?

F ecosystem, organism

G organism, organ system

H circulatory system, heart

J organ system, nervous system

Copyright © Holt McDougal. All rights reserved.

53. Look at the figure below. What is the primary function of the cells that make up the majority of cells in this organ system?

A carrying oxygen

B sending electrical signals

C covering and protecting other cells

D contracting and relaxing to produce movement

54. Where is <u>most</u> of the ATP produced in cells?

F the endoplasmic reticulum

G the ribosomes

H the mitochondria

J the Golgi complex

Copyright © Holt McDougal. All rights reserved.

55. A community of plants growing in a terrarium can grow even if the terrarium is sealed. Which of these possible explanations is not a reasonable hypothesis to explain how this can happen?

A The plants have adapted to using oxygen instead of carbon dioxide for photosynthesis.

B There was enough carbon dioxide initially to support the growth for a long time.

C Some kind of decomposer that releases carbon dioxide is growing in the terrarium.

D The sealed terrarium actually has small openings that allow exchange of air with the outside environment.

56. A scientist is studying a terrarium in which plants are growing, even though the terrarium is sealed. He hypothesizes that there must be some kind of decomposer that releases carbon dioxide that is growing in the terrarium. What experiment would not support or disprove this hypothesis.

F Sterilize the soil before planting a second terrarium.

G Use soil from a different source in a second terrarium.

H Add decomposers to the terrarium to determine whether any changes occur.

J Examine the soil from the terrarium through a microscope to look for evidence of whether any microorganisms live in the soil.

Copyright © Holt McDougal. All rights reserved.

57. **The graph below shows the data collected by a student as she watched a squirrel running on the ground. Use the graph below to answer the question that follows.**

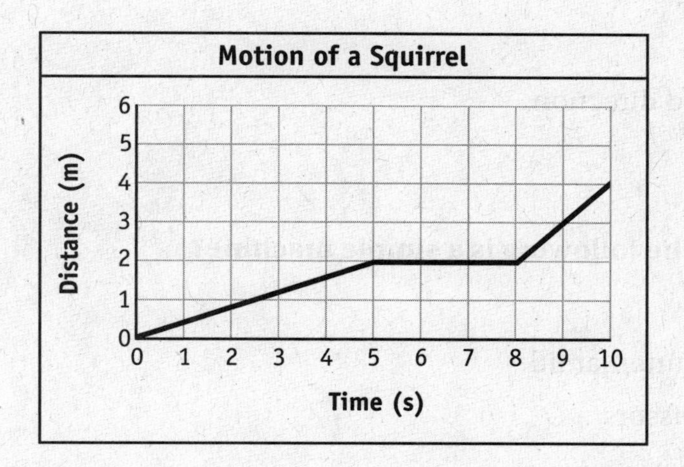

Which of the following best describes the motion of the squirrel between 5 s and 8 s?

A The squirrel's speed increased.

B The squirrel's speed decreased.

C The squirrel's speed did not change.

D The squirrel moved backward.

58. **A cell membrane is composed of which one of the following?**

F lipids

G nucleic acids

H phospholipids

J proteins

Copyright © Holt McDougal. All rights reserved.

59. An unbalanced force will always cause a change in an object's

A speed.

B direction.

C velocity.

D speed and direction.

60. Which of the following is a simple machine?

F bicycle

G peanut butter jar lid

H pair of scissors

J can opener

Copyright © Holt McDougal. All rights reserved.

61. Which statement below describes the mechanical advantage of the lever in the illustration?

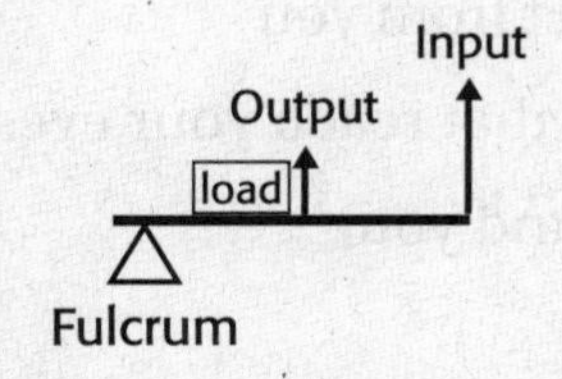

Second-Class Lever

A Mechanical advantage is less than 1.

B Mechanical advantage is equal to 1.

C Mechanical advantage is greater than 1.

D Mechanical advantage cannot be determined from the illustration.

62. In a multicellular organism each cell performs

F all functions.

G specific functions.

H random functions.

J the function of any cell it touches.

Copyright © Holt McDougal. All rights reserved.

63. What determines the color that your eyes see when you look at an object?

A the color of the light

B the distance of the object from you

C the wavelengths of light that reach your eyes

D the amount of light around you

64. Which of the following statements about photosynthesis is not true?

F It requires light energy.

G It takes place in the leaves of plants.

H It requires a pigment known as chromatid.

J It converts light energy into chemical energy.

Copyright © Holt McDougal. All rights reserved.

65. A child will be male if

A the sperm cell contains an X chromosome.

B the egg cell contains an X chromosome.

C the sperm cell contains a Y chromosome.

D the egg cell contains a Y chromosome.

66. During constructive interference of two waves,

F the amplitude increases.

G the frequency decreases.

H the wave speed increases.

J All of the above

Copyright © Holt McDougal. All rights reserved.

67. Which of the following best explains why levers usually have a greater mechanical efficiency than other simple machines?

A Levers have multiple applications.

B Levers are easy to manipulate.

C Levers are better at exploiting inertia than other simple machines.

D Levers tend to generate less friction than other machines.

68. A machine that has a mechanical advantage of less than one

F increases distance.

G multiplies force.

H increases output force.

J reduces distance and speed.

Copyright © Holt McDougal. All rights reserved.

69. On the map, note the two active volcanoes located within the Pacific Ocean that are not near a plate boundary. These volcanoes are most likely caused by

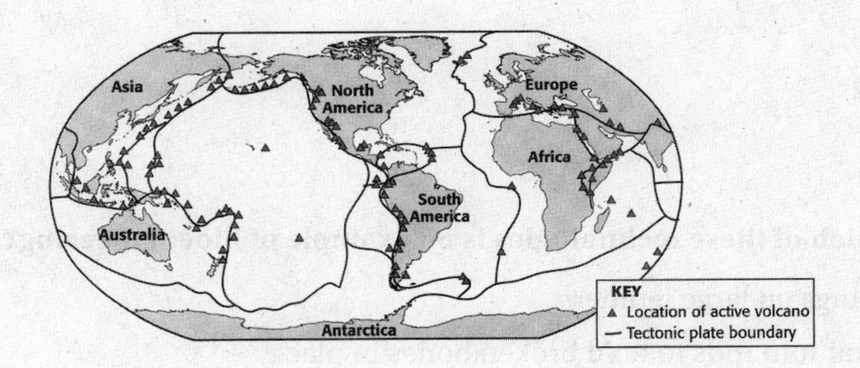

A hot spots.

B subduction zones.

C mid-ocean ridges.

D magma filling up gaps caused by a rift zone.

70. When you determine the streak of a mineral, what do you look at?

F oxidized particles of the mineral

G the product of the reaction with the streak plate

H tiny mineral crystals

J the impurities from the surface of the mineral

Copyright © Holt McDougal. All rights reserved.

71. How many chromosomes does a human egg cell contain?

A 23

B 32

C 46

D 92

72. Which of these technologies is an example of bioengineering?

F wings on large jetliners

G titanium rods to hold broken bones in place

H buses that run on hydrogen fuel

J a racing bicycle

Copyright © Holt McDougal. All rights reserved.

73. What part of the flower shown below is considered to be a male sex organ?

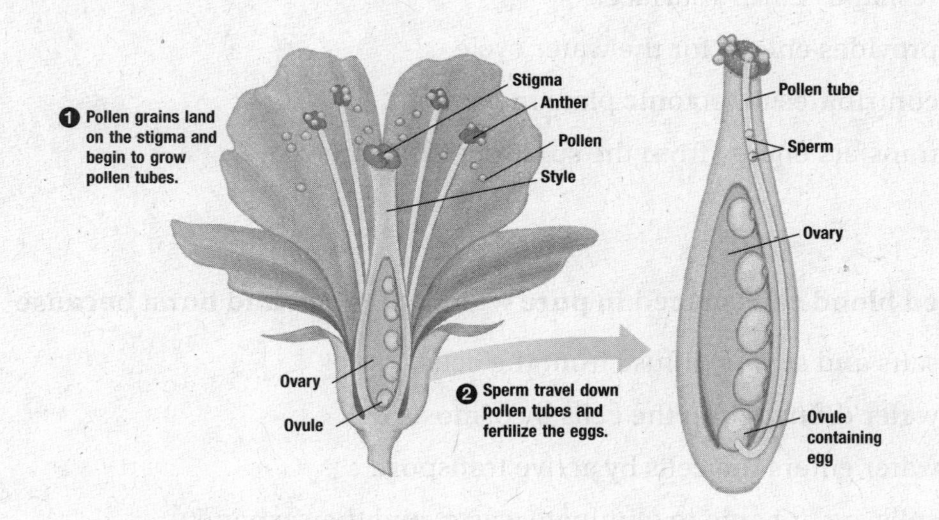

A ovary

B ovule

C stigma

D anther

74. Mountains and other major landforms on Earth result from forces generated by

F the separation, collision, and movement of tectonic plates.

G high winds during sandstorms.

H waves crashing against the coast.

J meteors striking Earth's surface.

Copyright © Holt McDougal. All rights reserved.

75. Weathering, transport, and deposition are forces that

- **A** reshapes Earth's surface.
- **B** provides energy for the water cycle.
- **C** contributes to tectonic plate movement.
- **D** transfers energy from the sun to Earth's surface.

76. Red blood cells placed in pure water will swell and burst because

- **F** salts and sugars diffuse from the cells.
- **G** water diffuses into the cells by osmosis.
- **H** water enters the cells by active transport.
- **J** cells are not able to eliminate waste, and they expand.

77. In multicellular organisms, specialization makes the organisms more efficient. Specialization means that

- **A** larger organisms are prey for fewer predators.
- **B** each type of cell has a particular job.
- **C** each species of organism has one goal.
- **D** the life span of organisms is limited to the life span of single cells.

Copyright © Holt McDougal. All rights reserved.

TCAP Test Preparation
Practice Test B

1. Look at the diagram below. If human blood did not contain component B, what would the blood be incapable of doing?

Components of Human Blood

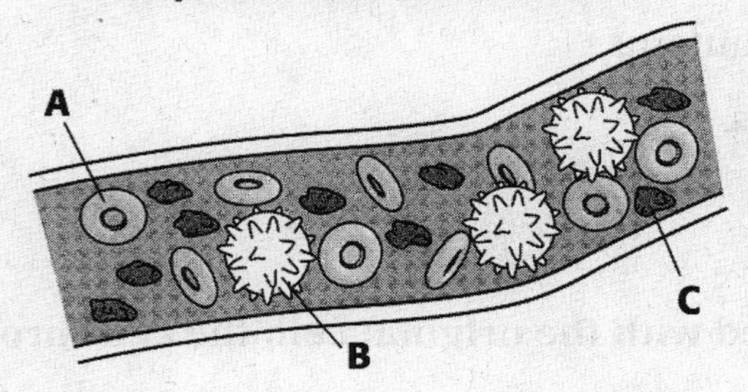

A clotting

B fighting disease

C carrying oxygen

D having red color

2. The heat in the interior of Earth increased as the planet formed, due to all of the following <u>except</u>

F collisions between planetesimals and Earth.

G radioactive material emitting energy.

H the motion of Earth around the sun.

J the force of gravity crushing the interior materials.

Copyright © Holt McDougal. All rights reserved.

3. Geological evidence of seafloor spreading includes

A magnetic reversal.

B distribution of plant and animal species.

C the Rocky Mountains.

D ocean currents.

4. When compared with the original cell, the cells produced during meiosis

F are identical to the original cell.

G have half the number of chromosomes.

H have doubled the number of chromosomes.

J lack any chromosomes.

Copyright © Holt McDougal. All rights reserved.

5. Which explanation, for the chart below, best describes why the author chose to present this data in a pie chart?

Distribution of Plants

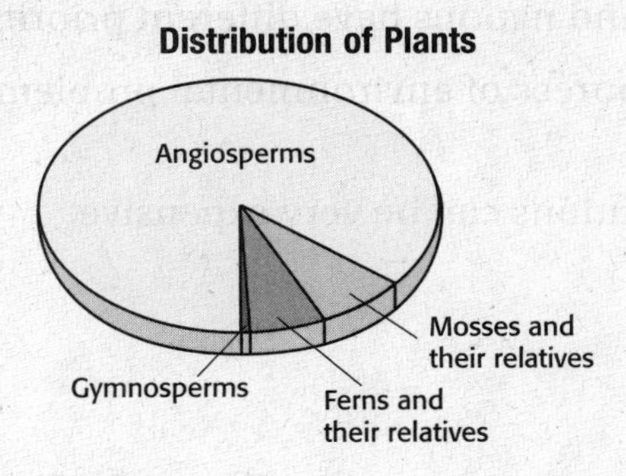

A She wanted to compare the percentages of each type of plant in an area.

B She wanted to show where different types of plants were located in an area.

C She wanted to demonstrate the continuous change of different plant populations over time.

D She wanted to show how different populations of plants had changed from one year to the next.

6. Gregor Mendel crossed a true-breeding tall plant (*TT*) with a true-breeding short plant (*tt*). What are the possible phenotypes for the offspring?

F all tall

G all short

H one tall, one short

J two tall, one short

Copyright © Holt McDougal. All rights reserved.

7. Solving environmental problems, such as air pollution, is sometimes a slow process because

A different people and nations have different priorities and needs.

B there are many sources of environmental problems, requiring different technologies.

C technological solutions can be very expensive.

D all of the above

8. Mitochondria are important organelles within a cell. What would most likely happen if a cell's mitochondria were not functioning properly?

F The cell's level of ATP would decrease.

G The cell's level of sugar would decrease.

H The cell would use lysosomes to release energy.

J The cell would create new mitochondria by cell division.

Copyright © Holt McDougal. All rights reserved.

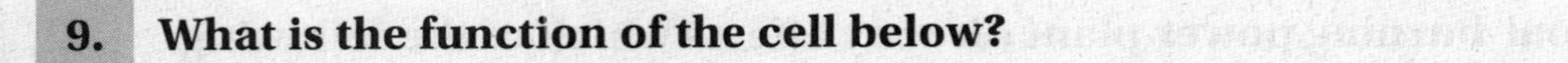

9. What is the function of the cell below?

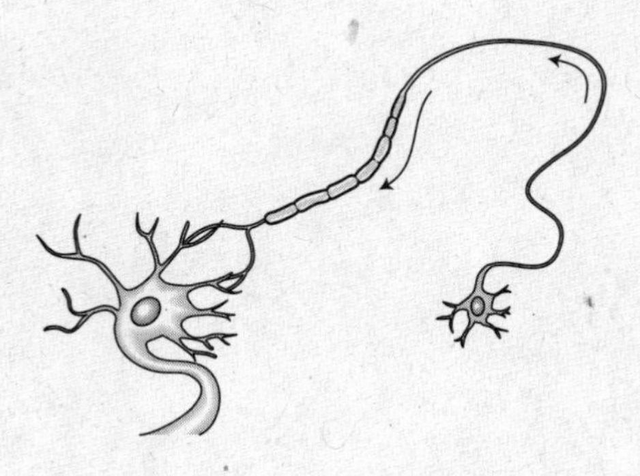

A to transmit signals

B to absorb nutrients

C to transport oxygen

D to contract and expand

10. According to the theory of plate tectonics, what process occurs at a divergent boundary?

F Two tectonic plates push into each other.

G Two tectonic plates slide past each other.

H Two tectonic plates move away from each other.

J One tectonic plate moves up and over another.

Copyright © Holt McDougal. All rights reserved.

11. A coal-burning power plant directly affects which of the following resources?

A air

B forests

C sun

D water

12. Which wave property is the same for all electromagnetic waves in a vacuum?

F speed

G pitch

H wavelength

J amplitude

Copyright © Holt McDougal. All rights reserved.

13. **David set up an experiment to determine how tall some house plants can grow. He tested the hypothesis that house plants grow taller when they are exposed to more hours of light. His experimental conditions are shown in the table below. The last column in the table shows how much the plants grew. Which factor is the variable?**

Experiment to Test How Tall Plants Grow

Group	Factors				Results
	Kind of Plant	Amount of Soil (ounces)	Amount of Water (cups per week)	Light exposure (hours per day)	Growth (cm)
1 (Control)	House plant	16	25	0	-8
2 (Experimental)	House plant	16	25	5	2
3 (Experimental)	House plant	16	25	10	15
4 (Experimental)	House plant	16	25	15	20

A kind of plant

B amount of water

C light exposure

D growth

14. **What kind of cells are plant cells?**

F multicells

G simple cells

H eukaryotic cells

J prokaryotic cells

Copyright © Holt McDougal. All rights reserved.

15. If two sound waves interfere constructively, you will hear

A a high-pitched sound.

B a softer sound.

C a louder sound.

D no change in the sound.

16. Newton's second law of motion explains why all objects fall with equal acceleration. Which of the following best summarizes this?

F Air resistance slows larger objects more than it slows smaller objects.

G Gravity exerts the same force on all objects, regardless of their size.

H The force-to-mass ratio is equal for all objects.

J Inertia is equal for all objects.

Copyright © Holt McDougal. All rights reserved.

17. Study the diagram of the muscle groups. The letter *A* points to a muscle that

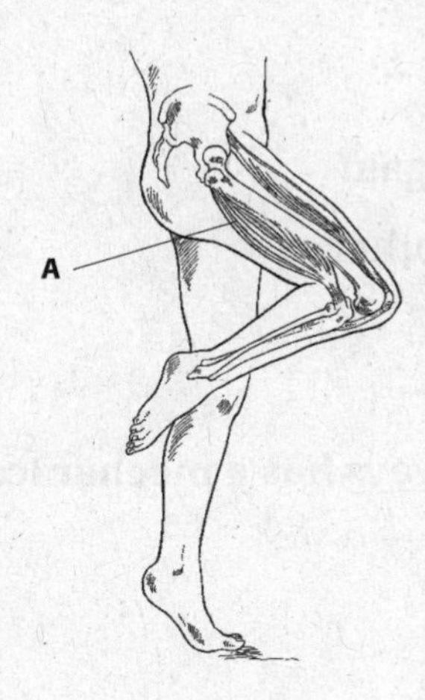

A bends part of your body and is called an extensor muscle.

B bends part of your body and is called a flexor muscle.

C straightens part of your body and is called an extensor muscle.

D straightens part of your body and is called a flexor muscle.

18. What two simple machines make up a pair of scissors?

F a lever and a wedge

G a lever and a wheel and axle

H a lever and a pulley

J a pulley and a wedge

Copyright © Holt McDougal. All rights reserved.

19. **Two scientists disagree about what the results of an experiment mean. What should they do?**

A run further tests

B study something else

C never work together again

D avoid talking to each other

20. **Which class of lever always has a mechanical advantage of greater than 1?**

F first-class

G second-class

H third-class

J none of the above

Copyright © Holt McDougal. All rights reserved.

21. During an experiment, why would a scientist record the exact volume of water in a laboratory fish tank?

A The fish tank might break, and he will have to buy a new one to replace it.

B The size of the fish tank might have an effect on the results of the experiment.

C By describing the tank, he prevents other scientists from copying his exact methods in their studies.

D If he reports the volume of the fish tank, he will not have to explain why his results did not match his hypothesis.

22. Which of the following is an example of a scientist's mathematical model?

F a Punnett square for a hybrid cross

G a description of prehistoric ecosystems

H a plastic version of the human skeleton

J a cartoon diagram of the structure of a cell

Copyright © Holt McDougal. All rights reserved.

23. The function of an organ is

A the way in which the organ is structured.

B unrelated to the organ's structure.

C the job that the organ performs.

D the arrangement of its parts.

24. Which of the following could cause a long-held theory to be challenged or even overturned by the scientific community?

F a popular celebrity who disagrees with the ideas outlined in the old theory

G new evidence that better matches the new theory than previous evidence

H a scientist who has strong beliefs that the old theory is wrong but little evidence

J a group of scientists who believe that new theories are usually better than old theories

Copyright © Holt McDougal. All rights reserved.

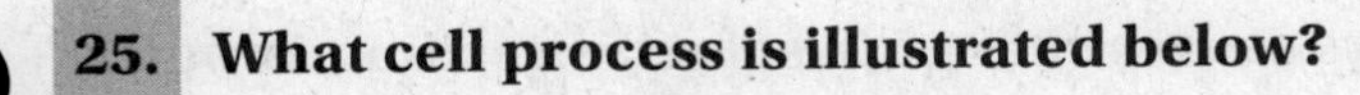

25. What cell process is illustrated below?

 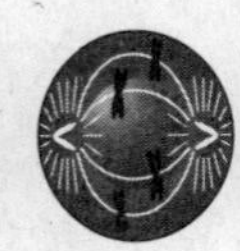 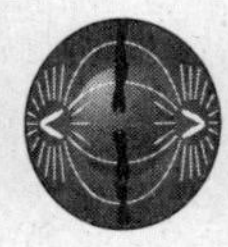  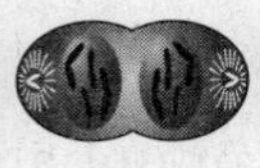 

A meiosis

B mitosis

C fertilization

D sexual reproduction

26. The sinking of Earth's crust can cause rock to melt. This molten rock can eventually produce

F a river.

G folded mountains.

H fault-block mountains.

J a volcanic eruption.

Copyright © Holt McDougal. All rights reserved.

27. Which of the following always causes change in speed, direction, or both?

A balanced forces

B unbalanced forces

C either balanced or unbalanced forces

D any combination of forces

28. Computers are often used in simulations to prepare a

F control variable.

G mathematical model.

H physical model.

J theory.

Copyright © Holt McDougal. All rights reserved.

29. Which statement below describes the mechanical advantage of the lever in the illustration?

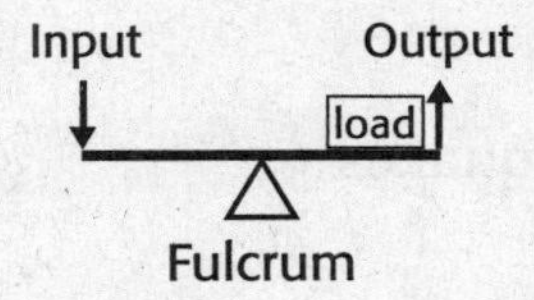

First-Class Lever

A Mechanical advantage is less than 1.

B Mechanical advantage is equal to 1.

C Mechanical advantage is greater than 1.

D Mechanical advantage does not apply to first-class levers.

30. Which of the following does not experience a change in velocity?

F A motorcyclist driving down a straight street applies the brakes.

G While maintaining the same speed and direction, an experimental car switches from gasoline to electric power.

H A baseball player running from first base to second base at 10 m/s comes to a stop in 1.5 seconds.

J A bus traveling at a constant speed turns a corner.

Copyright © Holt McDougal. All rights reserved.

31. What are the forces at work during the process of erosion?

A displacement, transportation, deposition

B deformation, faulting

C wind, water, ice, gravity

D volcanic eruption, earthquakes

32. A sideways force is applied to a block sitting on a table, but the block does not move. Which force prevents the block from moving?

F friction

G gravity

H weight

J surface smoothness

Copyright © Holt McDougal. All rights reserved.

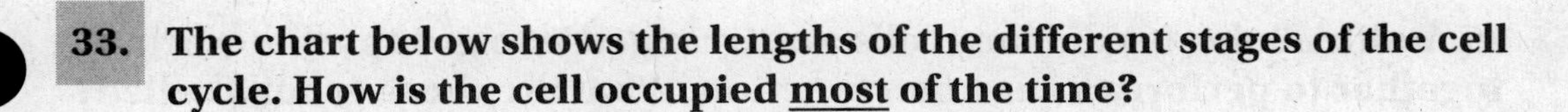

33. The chart below shows the lengths of the different stages of the cell cycle. How is the cell occupied most of the time?

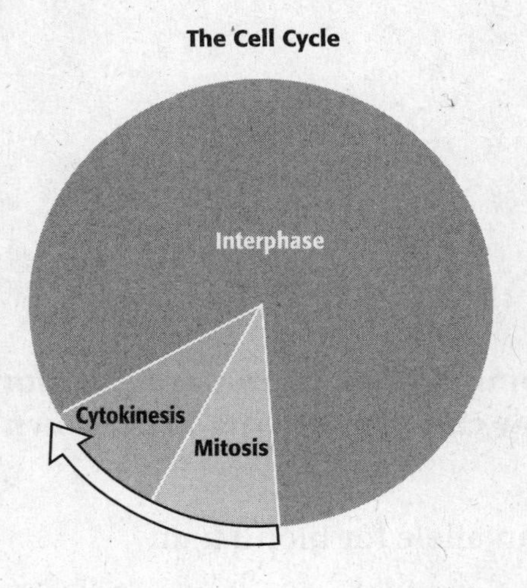

A copying DNA

B separating chromatids

C splitting into two daughter cells

D dissolving the nuclear membrane

34. What is the name of a cell vesicle that contains enzymes that destroy damaged cell organelles?

F endoplasmic reticulum

G Golgi complex

H lysosome

J mitochondria

Copyright © Holt McDougal. All rights reserved.

35. In multicellular organisms, cells are arranged into groups that work together to perform a common function. What are these groups called?

A systems

B bones

C tissues

D joints

36. Brown hair is a dominant trait compared to blond hair, which is a recessive trait. How can two parents with brown hair have a child with blond hair?

F Each parent has an allele for blond hair.

G One or both of the parents have an allele for brown hair.

H There was a mutation in the genes of one of the parents.

J This event cannot happen.

Copyright © Holt McDougal. All rights reserved.

37. Sharice made the sketch shown during a field investigation of Earth's structure. According to the sketch, which of the layers shown cannot be classified into the same group as the other layers according to its physical state?

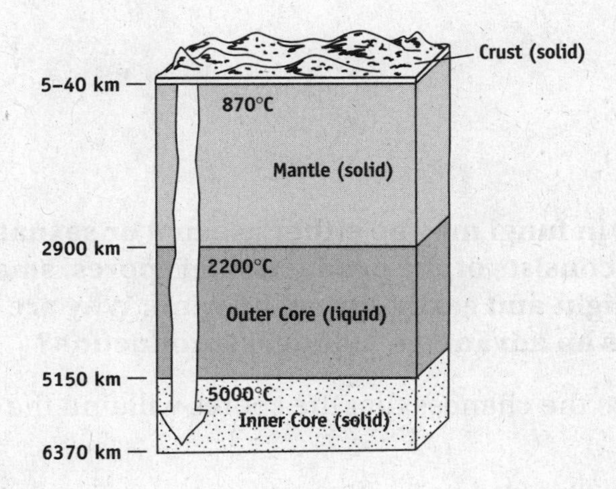

A inner core

B outer core

C mantle

D crust

38. According to Newton's third law of motion, if an ice skater exerts a force on a wall, the

F wall exerts an equal and opposite force on the skater.

G acceleration of the wall depends on the magnitude of the force.

H wall will not move because of its inertia.

J momentum of the skater is no longer conserved.

Copyright © Holt McDougal. All rights reserved.

39. Where are the instructions for traits found?

A meiosis

B genes

C mitosis

D Golgi bodies

40. Reproduction in fungi may be either asexual or sexual. Asexual reproduction consists of the production of spores, small reproductive cells that are light and easily spread by wind. Why are these characteristics an advantage to fungi reproduction?

F They decrease the chances that the spores will find the right growing conditions.

G Spores are small and can therefore reproduce more quickly.

H They decrease the chances that the spores will grow quickly.

J They increase the chances that the spores will find conditions suited to their growth.

Copyright © Holt McDougal. All rights reserved.

41. How does the third-class lever make work easier?

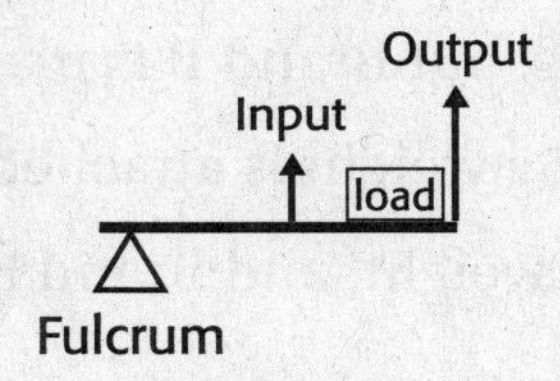

Third-Class Lever

A by changing the direction of the force

B by increasing both force and distance

C by increasing force and decreasing distance

D by decreasing force and increasing distance

42. Plants contain three types of tissue, each performing a specific function. These types of tissue are

F transport tissue, protective tissue, and ground tissue.

G roots, leaves, and stem.

H chloroplasts, membranes, and cytoplasm.

J connective tissue, protective tissue, and growth tissue.

Copyright © Holt McDougal. All rights reserved.

43. Anna wants to determine which brand of paper towel is stronger. Which of the following observations provides useful data?

A Brand A has blue stripes, but Brand B is plain white.

B Brand A tears when a 25g weight is attached, but Brand B does not.

C Brand A supports a 25g weight, and Brand B supports a melting ice cube that weighs the same.

D Brand A supports a large metal ball, but Brand B does not support a metal block of the same weight.

44. What is the outermost layer of Earth called?

F core

G lithosphere

H asthenosphere

J mesosphere

Copyright © Holt McDougal. All rights reserved.

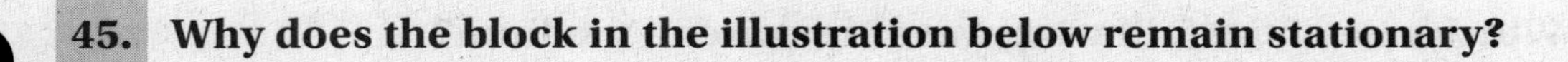

45. Why does the block in the illustration below remain stationary?

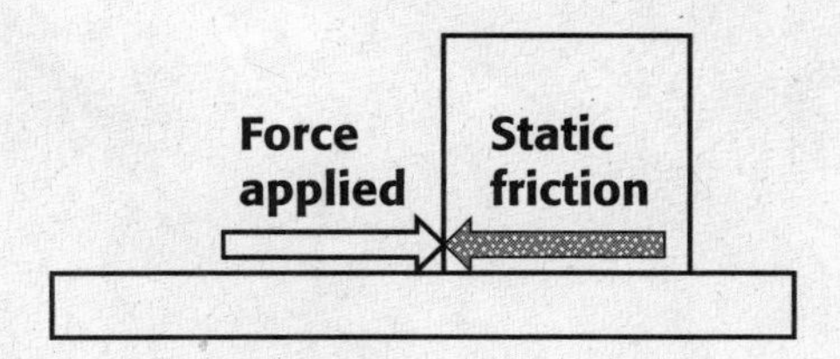

A No force is applied to it.

B The forces are balanced.

C Friction is stronger than the applied force.

D Gravity holds it in place

46. Which of the following is <u>not</u> a reason scientists conduct investigations?

F to explore new phenomena

G to confirm previous results

H to test how popular a theory is

J to compare competing theories

Copyright © Holt McDougal. All rights reserved.

47. Which one of the following waves requires a medium?

A electromagnetic waves

B mechanical waves

C light waves

D X rays

48. An object moving in a straight line with a constant speed has no unbalanced forces acting on it. How will the object's motion change over time?

F The object will gradually slow down and come to a stop.

G Centripetal force will cause the object to go into a circular orbit.

H The object's motion will remain unchanged.

J The object will move in a direction opposite the applied force.

Copyright © Holt McDougal. All rights reserved.

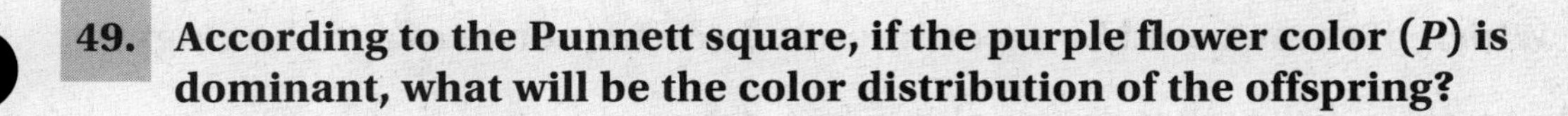

49. **According to the Punnett square, if the purple flower color (*P*) is dominant, what will be the color distribution of the offspring?**

	p	*p*
P	***Pp***	***Pp***
P	***Pp***	***Pp***

A all purple
B all white
C same number of purple and white
D mix of purple, white, and shades in between

50. **What can influence traits besides genes?**

F height
G albinism
H environment
J phenotype

Copyright © Holt McDougal. All rights reserved.

51. Which of these could be an unintended consequence of a technology?

A ceramic dishes that do not chip when dropped

B damage to forests by acid rain caused by automobile exhaust

C better crop yields due to irrigation using water from dam

D shorter travel times using a high-speed rail system

52. The greatest earthquake damage happens at the

F focus.

G boundary between tectonic plates.

H epicenter.

J seismograph station.

Copyright © Holt McDougal. All rights reserved.

61. Examine the illustration depicting the formation of the Hawaiian Islands. One of the active volcanoes on the island of Hawaii is Kilauea. If Kilauea is a shield volcano, which of the following conclusions is valid?

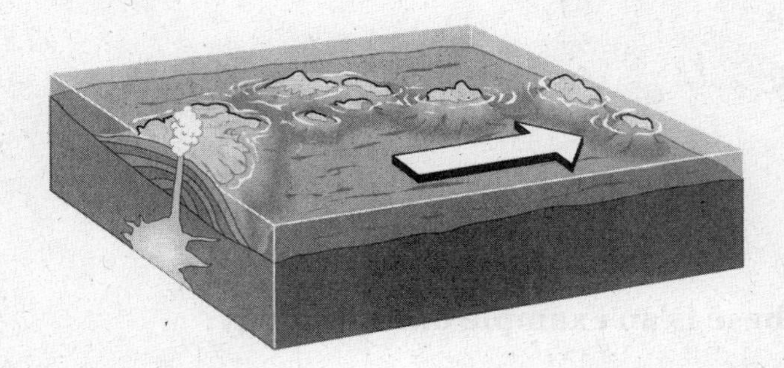

A Kilauea formed from repeated eruptions of low-viscosity lava that has spread over a wide area.

B Kilauea formed from repeated moderately explosive eruptions of pyroclastic material.

C Kilauea is primarily made up of alternating layers of lava and pyroclastic material.

D Kilauea has formed a cinder cone through repeated eruptions of high-viscosity lava.

62. Transport tissue is found only in

F animals.

G plants.

H one-celled organisms.

J nerves.

Copyright © Holt McDougal. All rights reserved.

63. The distance between any two adjacent crests or compressions in a series of waves is called

- **A** wavelength.
- **B** amplitude.
- **C** frequency.
- **D** speed.

64. Which of these is an example of technology?

- **F** boiling point of water
- **G** global climate change
- **H** Doppler radar
- **J** sound waves

Copyright © Holt McDougal. All rights reserved.

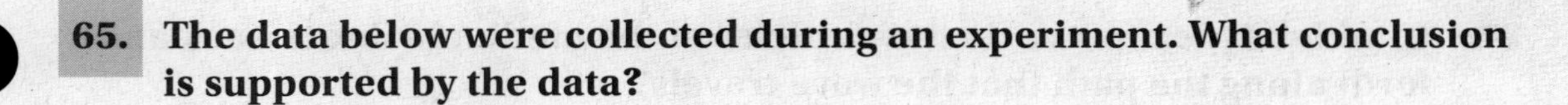

65. The data below were collected during an experiment. What conclusion is supported by the data?

Temperature (°C)	Time to double bacteria population (min)
10	130
20	60
30	29
40	19
50	no growth

A Bacteria always grow faster at lower temperatures.

B The rate of growth of this bacterial species increases as temperature is increased from 10° C to 50° C.

C Temperature has no effect on the rate of bacteria growth.

D Bacteria do not like higher temperatures.

66. What was the source of the original oxygen in the atmosphere about 3 billion years ago?

F volcanoes

G oceans

H cyanobacteria

J green plants

Copyright © Holt McDougal. All rights reserved.

67. In which type of wave do the particles of the medium vibrate back and forth along the path that the wave travels?

A longitudinal wave

B transverse wave

C mechanical wave

D light wave

68. A doorknob is what type of simple machine?

F lever

G screw

H wheel and axle

J pulley

Copyright © Holt McDougal. All rights reserved.

69. If the input force on this block and tackle system is 40 N, what is the output force?

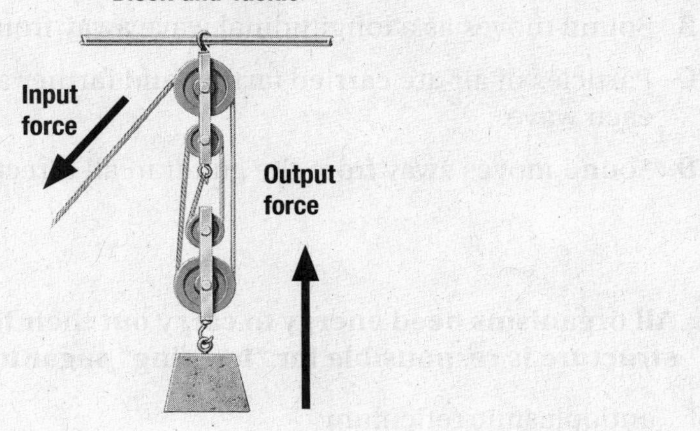

A 10 N

B 40 N

C 80 N

D 160 N

70. Waves, in which the particles vibrate with an up-and-down motion, are called

F sound waves.

G longitudinal waves.

H transverse waves.

J electromagnetic waves.

Copyright © Holt McDougal. All rights reserved.

71. Which of the following is not a true statement about what happens when a guitar is played?

A The strings of the guitar are vibrating.

B Sound moves as a longitudinal wave away from the guitar.

C Particles of air are carried farther and farther away from the guitar with each wave.

D Sound moves away from the guitar in all directions.

72. All organisms need energy to carry out their life processes. Which cell structure is responsible for "burning" sugar to release energy?

F endoplasmic reticulum

G mitochondrion

H nucleus

J ribosome

Copyright © Holt McDougal. All rights reserved.

73. Based on this diagram of the rock cycle, uplift and erosion are processes that form which of the following types of rock?

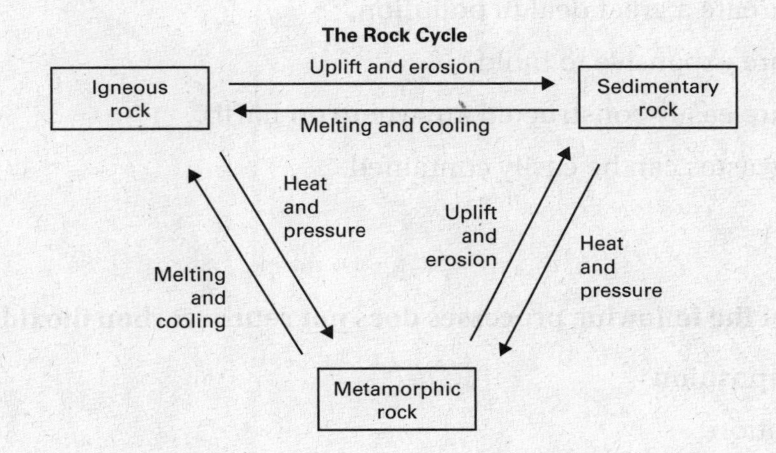

A metamorphic

B sedimentary

C igneous

D volcanic

74. A net force on a nonmoving object will cause the object to

F move away from the direction of the net force.

G move in the direction of the net force.

H reverse the direction of its movement.

J reduce the speed of its movement.

Copyright © Holt McDougal. All rights reserved.

75. Which of the following is a disadvantage associated with using solar energy panels to create electricity?

A They create a great deal of pollution.

B They are expensive to build.

C They are easily constructed anywhere on Earth.

D Their wastes can be easily contained.

76. Which of the following processes does <u>not</u> return carbon dioxide to the air?

A decomposition

B respiration

C combustion

D photosynthesis

77. What observation about the mid-ocean ridge shown in the figure below was used to support Alfred Wegener's hypothesis of continental drift?

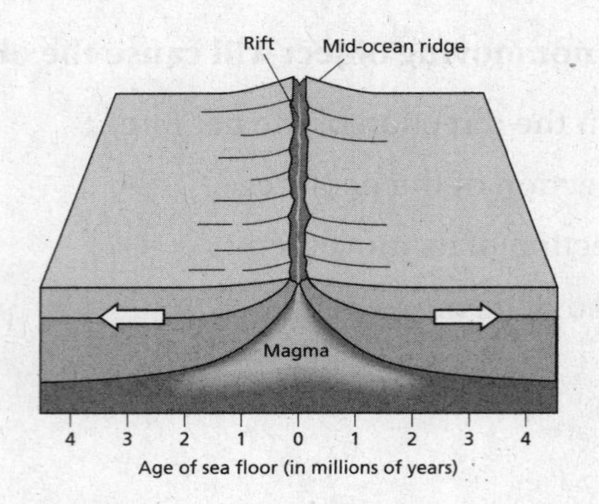

A Magma is very close to the surface at the rift.

B Sediments become thicker with increasing distance from the rift.

C Energy from the rift affects currents in the water of the Atlantic Ocean.

D The crusts on opposite sides of the rift appear to be moving towards one another.

Copyright © Holt McDougal. All rights reserved.